ESD Protection Methodologies

Energy Management in Embedded Systems Set

coordinated by
Maryline Chetto

ESD Protection Methodologies

From Component to System

Marise Bafleur
Fabrice Caignet
Nicolas Nolhier

ELSEVIER

First published 2017 in Great Britain and the United States by ISTE Press Ltd and Elsevier Ltd

ISTE Press Ltd
27-37 St George's Road
London SW19 4EU
UK

www.iste.co.uk

Elsevier Ltd
The Boulevard, Langford Lane
Kidlington, Oxford, OX5 1GB
UK

www.elsevier.com

For information on all our publications visit our website at http://store.elsevier.com/

British Library Cataloguing-in-Publication Data
A CIP record for this book is available from the British Library
Library of Congress Cataloging in Publication Data
A catalog record for this book is available from the Library of Congress
ISBN 978-1-78548-122-2

Printed and bound in the UK and US

Contents

Foreword 1

Our societies are going through a genuine digital revolution, known as the Internet of Things (IoT). This is characterized by increased connectivity between all kinds of electronic devices, from surveillance cameras to medical implants. With the coming together of the worlds of computing and communication, embedded computing now stretches across all sectors, public and industrial. The IoT has begun to transform our daily life and our professional environment by enabling better medical care, better security for property and better company productivity. However, it will inevitably lead to environmental upheaval, and this is something that researchers and technologists will have to take into account in order to devise new materials and processes.

Thus, with the rise in the number of connected devices and sensors, and the increasingly large amounts of data being processed and transferred, demand for energy will also increase. However, climate change is placing an enormous amount of pressure on organizations to adopt strategies and techniques at all levels that prioritize the protection of our environment and look to find optimal methods of using the energy available on our planet.

Over the past few years, the main challenge facing R&D has been what we have now come to refer to as green electronics/computing: in other words, the need to promote technological solutions that are energy-efficient and that respect the environment.

The set of books entitled *Energy Management in Embedded Systems* has been written in order to address this concern:

In their volume *Energy Autonomy of Batteryless and Wireless Embedded Systems,* Jean-Marie Dilhac and Vincent Boitier consider the question of the energy autonomy of embedded electronic systems, where the classical solution of the electrochemical storage of energy is replaced by the harvesting of ambient energy. Without limiting the comprehensiveness of their work, the authors draw on their experience in the world of aeronautics in order to illustrate the concepts explored.

The volume *ESD Protection Methodologies,* by Marise Bafleur, Fabrice Caignet and Nicolas Nolhier, puts forward a synthesis of approaches for the protection of electronic systems in relation to electronic discharges (ElectroStatic Discharge or ESD), which is one of the biggest issues with the durability and reliability of new technology. Illustrated by real case studies, the protection methodologies described highlight the benefit of a global approach, from the individual components to the system itself. The tools that are crucial for developing protective structures, including the specific techniques for electrical characterization and detecting faults as well as predictive simulation models, are also featured.

Maryline Chetto and Audrey Queudet present a volume entitled *Energy Autonomy of Real-Time Systems*. This deals with the small, real-time, wireless sensor systems capable of taking their energy from the surrounding environment. Firstly, the volume presents a summary of the fundamentals of real-time computing. It introduces the reader to the specifics of so-called *autonomous* systems that must be able to dynamically adapt their energy consumption to avoid shortages, while respecting their individual time restrictions. Real-time sequencing, which is vital in this particular context, is also described.

The volume entitled *Flash Memory Integration* by Jalil Boukhobza and Pierre Olivier attempts to highlight what is currently the most commonly used storage technology in the field of embedded systems. It features a description of how this technology is integrated into current systems and how it acts from the point of view of performance and energy consumption. The authors also examine how the energy consumption and the performance of a system are characterized at the software level (applications, operating system) as well as the material level (flash memory, main memory and CPU).

Maryline CHETTO

Foreword 2

Why this book? Why use it as a reference when seeking methods of protection against electrostatic discharge issues?

The truth is, what we believe to know about these devilish electrostatic discharges is not enough, proof of which are the many expert reports that include terms such as "ESD type fault". Is this likely to be satisfactory for anyone who has just designed, built or used a device that they thought was protected against this phenomenon by supposedly state-of-the-art methods?

Nonetheless, the highest-level precautions, such as those used in space-related devices, with their extreme quality and reliability requirements, are not enough to fully protect from these problems. Just like in other industrial areas, too many reports contain the infamous "ESD type fault" term.

Of course, prevention is better than cure; but in both cases, what must be done to avoid being faced (again) with the same diagnosis of a failure caused by electrostatic discharge? This is a key demand of designers, manufacturers and users, who, for once, share the same interest.

One option is to overprotect, build a Maginot line against electrostatic discharge. However, beyond the extreme costs, history has shown that such solutions are severely limited. Must we leave it up to chance, then? What can be done to avoid ESD (Electronic System Destruction) by ESD (Electro Static Discharge)?

A good first step is to read this book, going back to the most important points in order to choose, size and join the adapted structures depending on your needs and your operating conditions.

However, we must remain humble. Despite all that can be learned, the enemy is a sly beast and challenges us on the field of any new technological development. As a result, we must make it our objective to always get to know it better. It is unlikely that this type of failure will ever be 100% eliminated, but its drastic reduction is of major importance for all, and this book will surely contribute toward this goal.

Most remarkable in this book is the competence and mastery of the subject that is shown, page after page, by the authors, as well as their didactic approach, the sheer amount of useful information, and the systematic way in which it is organized. On that, enjoy the read, and down with ESD!

Philippe PERDU
CNES
Toulouse
France

Preface

When the authors of this book decided to start work at the LAAS-CNRS on the issue of failures caused by electrostatic discharge (ESD), the research team that they belonged to was working on improving performance levels and the robustness of MOS, IGBT and bipolar power electronic devices. It was our historical industrial partner, Motorola Semiconductors[1] that consulted us in 1996 regarding this issue, which they had encountered with their famous smart power technology, SMARTMOS™. By taking a closer look at the problems that the designers of these protections were confronted with, we quickly realized that our expertize on the physics of electronic power components would be very relevant, as, ultimately, the main requirement for these power structures is to dissipate the energy of the discharges over a small silicon surface. As a result, there is very little margin available for the design, and the operating conditions of the product are extreme (strong current densities, strong electric fields) and close to performance limits.

At the time, the LAAS-CNRS had also heavily invested in finite element simulation tools and even developed a tool for the calculation of the static voltage handling of power components, POWER 2D[2]. These tools provided a huge array of possibilities for understanding complex physical mechanisms such as access to information that was not available using the electrical test. We therefore accepted the challenge from Motorola and proposed an extensive use of physical simulations in order to solve it. This marked the start of our first thesis in the area of ESD, by Christelle Delage, which was followed by a number of other theses in collaboration with other industrial and academic partners. The results presented in this book are collated from these different theses, and we would like to thank all of the students

---

1 Motorola Semiconductors became Freescale Semiconductors in 2004, and then NXP Semiconductors in 2016.
2 [NOL 94].

who contributed toward improving our understanding of this domain: Christelle Delage, Géraldine Bertrand, Patrice Besse, David Trémouilles, Christophe Salamero, Nicolas Guitard, Amaury Gendron, Nicolas Lacrampe, Yuan Gao, Jason Jiyu Ruan, Johan Bourgeat, Frank Jezequel, Nicolas Monnereau, Marianne Diatta, Antoine Delmas, Houssam Arbess, Sandra Giraldo, Bertrand Courivaud, Rémi Bèges and Fabien Escudié. It is an interesting point that most of these students have gone on to become ESD experts in different companies all over the world (NXP France & Netherlands, ON Semiconductor France, STMicroelectronics France, GLOBALFOUNDRIES USA, AMS Austria and SERMA INGENIERIE France). We would also like to thank our industrial and academic partners without whom these studies would not have taken place: Motorola Semiconductors/Freescale Semiconductors, with which the LAAS-CNRS has had three successive shared laboratories from 1995 to 2008, and in which ESD was one of the research project themes, ON Semiconductors, STMicroelectronics, IPDIA, VALEO, CNES, the *Direction Générale de l'Armement* (DGA) and the AMPERE laboratory in Lyon and the IMS laboratory in Bordeaux.

In 2003, the LAAS-CNRS integrated the working group EOS-ESD into the failure analysis association ANADEF[3]. In addition to members of the semiconductor industry, many equipment manufacturers (automobile, aeronautics, military and space) were also members. These manufacturers made us aware of the issue of ESD robustness in electronic boards, as well as the lack of tools and methods for improving and characterizing it. As a result, the DGA, which was the leader of this working group at the time, suggested in 2004 that we begin a thesis on the modeling of the robustness of electronic boards regarding electric (EOS) and electrostatic (ESD) stress present in an environment of defense (Nicolas Lacrampe's thesis). We were pioneers in the domain, and our first publications at the EOS/ESD Symposium proposing a convergence of approaches for modeling and characterizing CEM and ESD were not always very well understood. Currently, the competence of LAAS-CNRS in the field of ESD systems is internationally recognized and Fabrice Caignet is the leader of a working group at the ESD Association[4].

Finally, we would like to state how much we have appreciated the richness and wealth of our interactions with the EOS/ESD community over the past twenty years during the various annual conferences and workshops. In particular, we would like to thank Steven Voldman for introducing us to the community in 1998; Philippe

---

3 ANADEF (http://www.anadef.org/): *founded in 2001, this French association groups together industry actors and scientists concerned with the mechanisms of failure of electronic components and assembly lines, with the goal of improving prevention, detection and analysis.*
4 ESDA (EOS/ESD Association) https://www.esda.org/.

Perdu for having linked us to the challenges of failure detection using laser stimulation; James Miller for taking us with him on the adventure of creating the IEW workshop; Chavarka Duvvury for his initiative in the ESD Academia committee and for trusting us to act as its leader, and Nate Peachey and Robert Ashton for supporting us in our role within the standardization committees of ESDA.

Throughout this book we have tried to transmit our convictions regarding protection approaches, which we believe must be tackled in a global fashion. During our studies, we have seen that protection against latch-up can greatly degrade the ESD protection of an integrated circuit, or that good immunity to electromagnetic interference obtained using a high value capacitor can degrade the ESD robustness of an electronic board.

Following an introduction to the phenomenon of electrostatic discharge and its effects on electronic components, in Chapter 1 we present the various normalized test techniques that are used to qualify the ESD robustness of a component or of an electronic board. These test techniques do not allow us to understand why some components fail to pass the qualification. To this aim, other characterization and defect localization techniques are required to put corrective measures into practice, which are presented in Chapter 2. In Chapter 3, we look at the various strategies' protection both at component and system levels. With the increasing complexity of integrated circuits, it is important to be able to provide a simulation in which the implemented ESD protection strategy provides the desired protection, while not harming the performance levels of the circuit. This aspect is covered in Chapter 4, where we detail the main features and difficulties related to the different types of simulation: finite elements, SPICE-type and behavioral ones. In Chapter 5, we present several study cases that illustrate the approaches described in the previous chapters. We finish with a summary of the most important rules used in ESD protection and with a general conclusion.

We hope that our experience will prove useful to new designers and failure analysis engineers. Happy reading!

Marise BAFLEUR
Fabrice CAIGNET
Nicolas NOLHIER
May 2017

Introduction

Failures caused by electrostatic discharges still make up a significant percentage of electronic components field returns. In some areas, such as the automotive industry, this percentage can get close to 20%. In this chapter, we shall first introduce the concept of an electrostatic discharge event by explaining its origins. Next, we shall describe the impact they have on the reliability and robustness of integrated circuits and of electronic systems. Finally, we shall discuss the precautions that must be taken in order to minimize the risks of failure during manufacturing and assembly of these components by establishing zones protected from these disturbances.

I.1. Origin of electrostatic discharge

Electrostatic discharge (ESD) is the result of a rapid, high-intensity transfer of charges between two objects of different electrostatic potentials [GRE 91]. This phenomenon of discharge is relatively common. The most spectacular example is that of lightning, which takes place following the accumulation of static electricity between storm clouds, or between one of these clouds and the ground. The electric potential difference between the two points can reach up to 10–20 million volts and produces plasma during discharge, resulting in an explosive expansion of the air by heat escape. As it dissipates, the plasma creates a lightning strike and thunder.

Although on a much smaller scale, a human body is electrically charged and discharges several times a day. For example, walking on a synthetic rug causes an accumulation of electrons in the body, and this can lead to an electric shock – the discharge – when a metallic door handle is then touched. This little shock frees the accumulated static electricity. The phenomenon, called triboelectricity, is caused by an initial unbalance of charges between two bodies [VIN 98]. Human beings begin to feel discharges when the charging voltage reaches approximately 3.4 kV. From 15 kV, the discharge begins to cause pain.

There are other ways of generating a charge unbalance, such as generation by induction or even contact with previously charged objects [VIN 98].

The various associated mechanisms of discharge create considerable electrostatic voltage over short duration and strong currents. Several studies have shown that the waveform of these discharges depends not only on the characteristics of the source and of the discharge circuits (surface of contact between two objects), but also on other parameters (relative humidity of the air, approach velocity of a charged body) [GRE 02].

Table I.1 provides some examples of the generation of electrostatic charge by triboelectricity and shows the considerable effect of air humidity on the level of the discharge.

Activities generating electrostatic electricity	Electrostatic voltage (kV)	
	10% relative humidity	55% relative humidity
Walking on a synthetic carpet	35	7.5
Walking on a vinyl floor	12	3
Removing the integrated circuit from a plastic tube	2	0.4
Removing an electronic board from bubble wrap packaging	26	7

Table I.1. *Examples of activities generating static electricity and the impact of air humidity on the associated level of electrostatic voltage*

ESDs belong to the family of electric overstress (EOS). An EOS is defined as the exposure of a component or system to a current or voltage level that is above its maximal specifications. Usually, this is a stress by surges whose conditions are low in amplitude (5–10 V), long in duration (1 μs to 10 ms) and of moderate current (100 mA to 1 A). The resulting energy can be several orders of magnitude greater than that of an ESD stress and can result in extensive damage to the oxide, metal and/or silicon.

ESD corresponds to a transfer of electrostatic charges between two objects or surfaces whose electrostatic potentials are different. This is a high-voltage event (1—10 kV) of high current (1–10 A) and is short in duration (1–100 ns). The energy of an ESD stress is in the order of a few micro-Joules, which can induce failure modes by local fusion of the silicon and/or breakdown of the gate oxides.

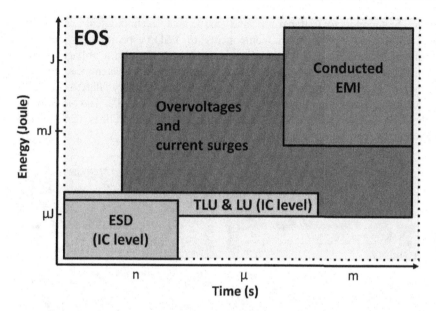

Figure I.1. *Graph illustrating the energy and frequency ranges of various EOS (Electrical Overstress) events. TLU and LU stand for transient latch-up and latch-up, respectively, and EMI for electromagnetic interference*

The graph in Figure I.1 illustrates the diversity of EOS events in terms of associated energy and of frequencies. ESDs are the most rapid (~GHz) and least energetic events. Next are the latch-up events, which correspond to the activation of the parasitic thyristor of CMOS technology in a static (LU) or dynamic (TLU) mode. In this graph, we have also reported conducted electromagnetic interferences (EMIs). This is an electric signal of undesired frequency that is superimposed over the useful signal. This parasitic signal can degrade equipment function. The source of electromagnetic emissions can be natural or artificial in origin, and intentional or unintentional. The disturbances shown in the graph are of low and medium frequencies, for a range of frequencies lower than 5 MHz, propagating mainly in a form conducted by the cables. Their duration can be of a few tens of ms. The conducted energy is significant and as a result, there is a risk of the materials being destroyed, in addition to malfunction.

I.2. Impact on the electronics

Catastrophic failures caused by ESD only started to be looked at very seriously with the appearance of microelectronic technology and the start of their widespread application in everyday life. More particularly, the invention of the MOS (Metal

Oxide Semiconductor) transistor and associated technological developments revealed the sensitivity of its components to ESD, especially its gate: some components could be destroyed during a transient ESD with a voltage as low as 10 V. In the 1970s, the failure of electronic components and systems caused by ESD started to increase exponentially. As a result, the military started to develop standards for testing the immunity of electronic products to ESD. The oldest of these is the MIL-STD-883E Method 3015.7 [MIL 89], which defines tests regarding discharge induced by a human body.

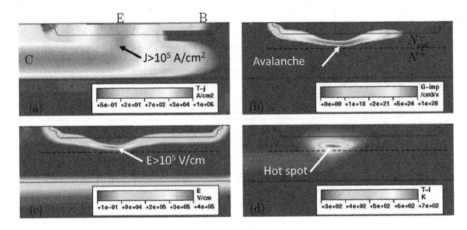

Figure I.2. *2D physical simulation of an ESD protection structure based on a bipolar NPN transistor: the different sections represent the current density (a), the electric field (c), the carrier generation rate by impact (b), and the temperature (d) [TRE 04a, TRE 04b]. For a color version of this figure, see www.iste.co.uk/bafleur/esd.zip*

This ESD can take place throughout the entire life of an electronic component, from its manufacturing, to its assembly on an electronic board, to finally its use in some type of application. They involve high current densities (> 10^5 A/cm^2) and very intense electric fields (> 10^5 V/cm), which can lead to failures. These current densities are directly dissipated through the silicon chip. This dissipation of power takes place within small volumes and results in a localized increase in temperature, which can lead to thermal damage, going as far as fusion of the material. Figure I.2 presents the results of a physical simulation of an ESD protection structure during discharge and is a good illustration of the extreme conditions in which these protection structures must operate.

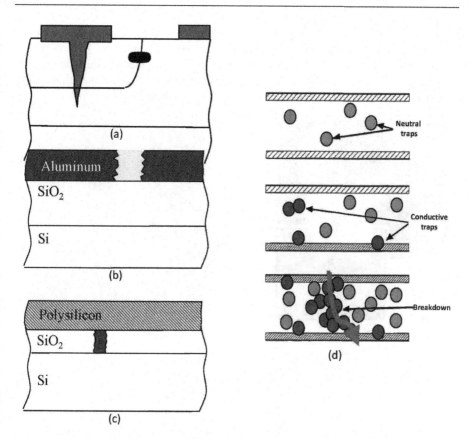

Figure I.3. *The different types of failure induced by an ESD: hot spot at a reverse-bias junction causing localized fusion (a), migration of the metal through the silicon under the effect of strong current densities (a), localized metal fusion (b), microfilament in an oxide gate under the effect of strong electric fields (c). The strong electric fields generate traps within the oxides that ultimately create a percolation path and the microfilament (d). For a color version of this figure, see www.iste.co.uk/bafleur/esd.zip*

The faults generated by an ESD are usually characterized by their small size (≤ 10 µm), except for stresses at the system level, which can be much larger. Figure I.3 illustrates the different types of faults that can be encountered, depending on their origin. In the presence of a strong electric field, at the level of a reverse-bias junction, a hot point can appear, which in some extreme conditions can reach the fusion temperature of silicon (1,414 °C) and lead to the local degradation of the junction by increasing its leakage current, or even by inducing a short circuit (a). The high current densities within metallizations such as aluminum (fusion point 660 °C) can lead to open circuits (b), or to migration of the metal within the silicon at the level of the contacts (a), which, when in the presence of thin junctions, can lead to an increase in the leakage current, or to a short-circuit.

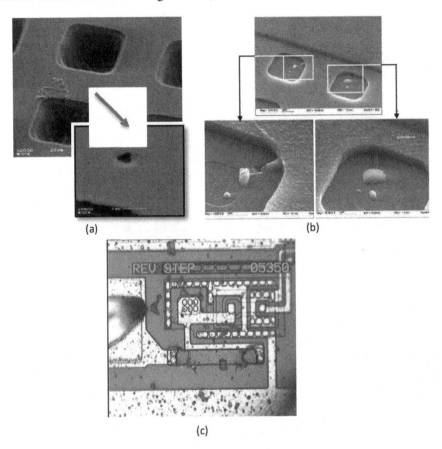

(a)   (b)

(c)

Figure I.4. *Examples of defects induced by an ESD: (a) breakdown of the gate oxide, (b) filamentation in the silicon at (c) fusion of a polysilicon resistor at both extremities. Photographs from [BAF 10]. For a color version of this figure, see www.iste.co.uk/bafleur/esd.zip*

Strong electrical fields, generated by ESD, result in high voltages across the circuits. In the case of MOS technology, this high voltage can appear at the terminals of a dielectric, such as the gate oxide of an MOS transistor. When the voltage of the oxide goes beyond the breakdown voltage level, it breaks and causes an irreversible failure by creating a microfilament through thermal heating (c). Usually, gate oxides can deal with electric fields of 6–10 MV/cm before breakdown occurs. Before breakdown, these strong electrical fields generate traps, which first induce a drift in the threshold voltage of the component, and ultimately, create a percolation conduction path (d).

Figure I.4 presents some examples of faults produced by different types of ESDs.

The development of microelectronics technology following the well-known Moore's Law [MOO 65] has resulted in an impressive decrease in the size of transistors, toward nanometric scales, increasing both integration density and performance yields. New production processes, such as ultrathin gate oxides using high permittivity dielectrics (HfO_2, for example), shallow junctions, highly doped drains, silicides, the use of new materials for interconnections (copper instead of aluminum), all make the circuits more sensitive to ESD. The volume into which the energy of the discharge is dissipated becomes smaller and smaller. The thinness of gate oxides means a smaller breakdown voltage. The interconnections are thinner and the dissipation energy is less efficient. Circuit failures are therefore happening at lower and lower levels of static voltage, as illustrated in Table I.2 from the provisions of the International Technology Roadmap for Semiconductors [ITR 05].

Year	2004	2007	2010	2013	2014	2016	2018	2020
Technological node	90 nm	65 nm	45 nm	32 nm	28 nm	22 nm	18 nm	14 nm
Max static charge (nC)	1	0.5	0.25	0.125	0.1	0.06	0.04	0.025
Max static voltage (V)	100	50	25	12.5	10	6	4	2.5

Table I.2. *Evolution of performance in terms of robustness to the static charges of CMOS technological nodes according to the ITRS 2005 [ITR 05]. Voltage is calculated for a component with an equivalent capacitance of 10 pF*

I.3. ESD Protected Area or "EPA"

In order to ensure the safe handling of electronic components, during their assembly and mounting onto electronic boards and into equipment with regard to ESD phenomena, most companies have established zones that are protected from ESD, also known as EPA (ESD Protected Area). In an EPA, the goal is to minimize the creation and accumulation of static charges.

To create an EPA, several measures must be put in place in order to define an environment that is safe from ESD phenomena. For this zone to be efficient both in terms of cost and performance yield, it must combine several prevention techniques, along with a precise working method.

In an EPA, the possibility of ESD events occurring must be minimized, and if such an event occurs despite these precautions, the associated charge must dissipate quickly without discharging throughout an electronic component.

These different measures can be split into several categories:

– *Environment control:* one of the key elements for limiting the production of the static charges involves the floor covering of the work area. It is essential that it be dissipative regarding static charges. This coating must have a resistance of less than 10^9 ohms but cannot be completely conductive for safety reasons. On top of this, the air can be controlled using humidifiers and air ionizers.

– *Use of antistatic products:* nowadays, there are several products that can be used in the handling of electronic components that have been made antistatic. For example: antistatic wrapping, conditioning rods for dissipative integrated circuits or any other wrapping material.

– *Workbench and antistatic auxiliary accessories:* the workbench surface must be dissipative and include antistatic wrist straps and chairs, so as to ensure the operator is always grounded (using a high value dissipative resistor for security reasons).

– *ESD compliant tools:* another important point involves all of the tools used in handling components. The tools, and especially soldering irons, can quite easily transfer charges to an electronic component through the metallic elements involved, which provide a low impedance pathway that can generate high current levels and the destruction of the component. An important point is to ensure the quality of the earthing of the equipment used and their good connection to the ground.

– *Antistatic clothing:* nowadays, clothes contain increasing amounts of synthetic fibers, liable to producing large amounts of static charges. It is therefore important to use antistatic clothing:

- An antistatic smock worn over normal clothes;

- Antistatic shoes, antistatic shoe covers or use of ESD heel-grounders (which play the same role as antistatic wristbands) on normal shoes.

– *Storage and transport:* It is rare for a component to be assembled in the same factory where it was manufactured. They must therefore be stored using the same measures used in working areas, by using dissipative materials:

   - Antistatic bags;

   - Shelves and cabinets that are ESD compliant.

– *Exclusion of any static charge generating materials:* polystyrene, bubble wrap, plastic glass or any other plastic object.

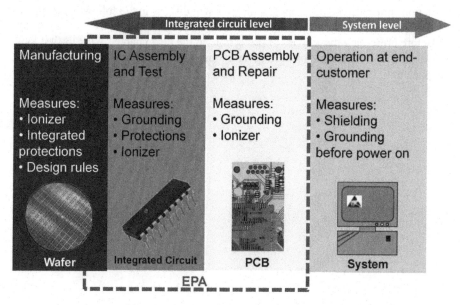

Figure I.5. *Main protective measures applied to the different steps of manufacturing, from the integrated circuit to the electronic equipment. The EPA zone is limited to the rectangle with dashed green lines. For a color version of this figure, see www.iste.co.uk/bafleur/esd.zip*

It is very important for all companies supplying components (integrated circuits or electronic boards) and electronic equipment repairers to have an EPA to protect themselves from ESD effects, which can immediately destroy components or create latent defects. Given that the component is not located in an EPA zone during use, it is also important to set up specific protections, which we shall talk about in the following chapters. Figure I.5 summarizes the main steps that are usually taken at the different steps of manufacturing, from the integrated circuit to the electronic equipment itself.

I.4. Conclusion

ESD remains a challenge in terms of the reliability and robustness of electronic components, partly due to the rapid development of technology towards nanometric sizes, and the appearance of a new technological node every 18 months. In order to counter the risk of failure during the manufacturing and assembly of components, a first step involves setting up zones that are protected from ESD. These steps, which involve controlling the environment and establishing precise handling practices (wearing of antistatic wrist straps, etc.) are not enough, and specific electronic protections must also be developed, which we will discuss in the following chapters.

Before presenting these types of protection, in the next chapter, we will first look at the different methods for testing and qualifying the robustness of an ESD component.

ESD Standards: From Component to System

In this chapter, we look at the different ESD standards that are used for qualifying integrated circuits (ICs) and electronic systems. At the component level, these standards provide a guarantee that the IC is able to survive the various steps of assembly. This means that power is not supplied to the IC whilst it undergoes these different stresses. At the system level, the electronic board must be able to resist various ESD stresses both when the supply is on and when it is off. This results in two possible types of failure: physical or hard failure, which causes the destruction of the component, and functional or soft failure, which causes a temporary malfunction of the circuit.

1.1. Standards: From component to system

Electrostatic discharges can have devastating consequences on the electronic systems that occupy a primary position in everyday life: multimedia, automobiles, aeronautics, etc. Such mobile electronic systems, also called embedded systems, must respond to operating constraints in real time. In the automotive and aeronautic industries, dependability of these systems is a priority in order to ensure user safety. No operating defects can be tolerated for these applications.

1.1.1. Increasingly sensitive electronics

From the 1990s to the 2000s, the level of robustness of integrated protection systems on chips dropped continuously, as shown in Figure 1.1 [EOS 16].

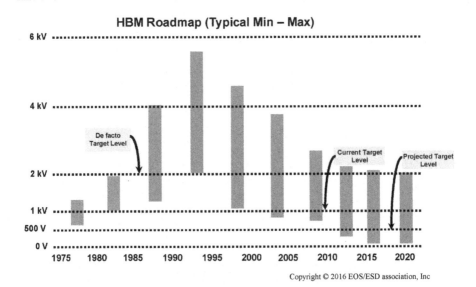

Figure 1.1. *Evolution of the robustness of electronic components to ESD (courtesy of the ESD Association) [EOS 16]*

This graph, published by ESDA (ESD Association) in the document entitled "2016 ESD Technology Roadmap" [EOS 16], shows the state of the HBM (Human Body Model, a standard that we shall explain in the following section) robustness required at the input and output of components. The levels of robustness can be seen increasing up to the 1990s and then decreasing quite significantly. For applications that use the more aggressive technologies (smaller sizes), this level is currently below 1.5 kV HBM. In EPA zones, it is now possible to guarantee a near absence of discharge in assembly areas. As ESD protections integrated into the circuit are initially sized to protect these circuits within these zones, it is no longer necessary to develop protection structures that are able to withstand high intensity discharges, especially as these are expensive in terms of their silicon footprint. As a result, since the 1990s, integrated protection systems are less robust.

However, in the last 10 years, with the rise of nomadic or embedded electronic systems, the level of robustness is no longer tied only to the constraints of the assembly area but also to recommendations from electronic manufacturers (EMs) who are looking to provide embedded electronic systems that are robust. C. Lippert from Audi automobiles presented, at the IEW 2010 workshop, an example of the requirements of automotive manufacturers proposing to shift the ESD constraint onto semiconductor manufacturers to limit failures of the system in the field [LIP 10].

Although these requirements relate to automotive applications, it is clear that at each step of system design, ESD must be considered as early as possible in the production line in order to avoid failure of the final system, which would be very costly. Nowadays the objective is "zero defects" throughout the life of the system.

Currently, the strategy for ESD protection at the system level is not well mastered. This is due to the fact that the issue is a new one and that EMs do not have any information on the components and the techniques used to protect them. The reliability constraints of products have also increased, making it harder to consider ESD during system design. Two types of failure can appear during ESD stress. The first is a physical failure, most often resulting in the destruction of the component (hard failure). The prevention methods used to manage security margins regarding ESD still display a considerable lack of certainty. When such a failure is detected, late changes to the system or to the ICs in order to solve it delay market release and are very costly. The second type of failure, which is a functional failure (called a soft failure), is linked to reversible malfunction, such as RESET or loss of function in the system. Analysis of this type of defect is far more complex and often requires in-depth knowledge of the system and ICs, as well as specific test methods.

The state of current research shows that there are gaps in terms of the methodology, models and tools used to simulate the hardware and functional robustness of an electronic application. Currently, the design of embedded systems is based on experience. There is a divide between integrated device manufacturers, who are not willing to provide information on the protection structures of the components, and system providers, who need models and characterization tools in order to develop robust systems. The approach presented in this book is an attempt to find an open ground for discussion between these two worlds.

1.2. Component level standards: HBM, MM, CDM, HMM

The protection levels of ICs are guaranteed by standards that specify the levels and stress waveforms. The robustness levels of components can be found in the technical documentation of manufacturers for different standards. Most of these standards have been developed in order to avoid the destruction of components in assembly zones. The HBM, CDM and MM standards are the most commonly used ones. These three models have been defined depending on the discharges produced:

– during contact with a charged human body (HBM mode – "Human Body Model") [ESD 01, ANS 14a];

– during contact with a charged metallic machine (MM model – "Machine Model") [ANS 99a];

– during earthing of a previously charged circuit (CDM model – "Charged Device Model") [ESD 99, ANS 15a].

The HBM model was the first to be developed, with standard MIL-STD-883D [MIL 89], and is still used as a reference in most industrial tests.

For this type of discharge, the model describes what happens when an already charged human (through movement, walking) discharges in an electronic circuit that he touches with his finger. During discharge, the component is assumed to have a lower potential than that of the human.

Another model used in the industry is the MM model. It was developed in Japan by the car industry as a "worst case" of the HBM model but is rarely used nowadays. This model represents the discharge of a charged object or a metallic machine coming into direct contact with an electronic circuit. The HBM and MM models describe the same discharge mechanism but with two different waveforms [BAR 04].

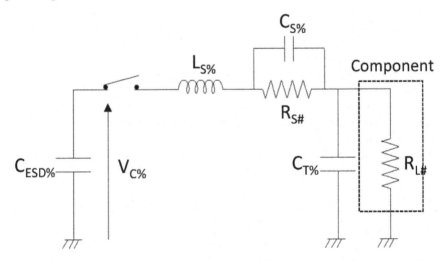

Figure 1.2. *Equivalent RLC diagram of the HBM, MM and CDM models*

These two models reproduce the electrostatic transfer of a charge into a device, while the CDM model reproduces the electrostatic transfer of a discharge from an initially charged device. The common link between these three models is the use of an RLC type circuit to simulate the discharge current that is caused by these different examples of ESD stress. Generally speaking, in these models the production of an ESD stress is carried out by the discharge of a previously charged electrical capacitance through a circuit under test. The circuit is modeled using a known impedance which is usually small.

Each standard specifies a waveform of the current injected into the product. For each of these, a tester is used who systematically specifies a charge capacitance and network, often a series resistance [HYA 02]. Parasitic elements (RLC) can be added to this very simple configuration to provide the dynamic behavior of the stress. Figure 1.2 shows how two waveforms can be generated from a shared electrical diagram. The HBM and MM models describe the same discharge mechanisms but with different waveforms [BAR 04] and different component values. The discharge is viewed as the closing of a switch between a pre-charged capacitance and the RLC + circuit under test, simultaneously representing the human/machine/circuit interface (C_{ESD}, R_S), the circuit itself (R_L) and the parasitic components of the tester, onto which model (L_S, C_S, C_T) is implanted. These values, which are fixed by the associated standards, are given in Table 1.1.

Model	C_{ESD} (pF)	R_S (Ω)	L_S (μH)	C_S (pF)	C_T (pF)
HBM	100	1500	5	< 5	< 30
MM	200	0	0.5/2.5	0	< 30
CDM	10	10	0.1	0	0

Table 1.1. *Characteristics of discharge models HBM, MM and CDM*

The waveforms of the three main standards at the component level are shown in Figure 1.3 for recommended charge levels. These standards, which are categorized in Table 1.2, define the various voltages and current levels that characterize the associated electrostatic discharge, as well as the methodology used to inject ESD stress into the circuit.

Depending on the domain of application and the voltage levels, each of the ESD models used (HBM, CDM and MM) in the test has its own classification system for differentiating circuits that are more affected by ESD from those that are less affected. For example, a circuit is considered to be robust with regard to ESD if it can survive a pre-charge voltage of V_{HBM}=4 kV or V_{CDM}=200 V. In the car industry higher specifications, such as voltages of more than 10 kV, are required to simulate the hostile environment in which the circuits must be able to operate.

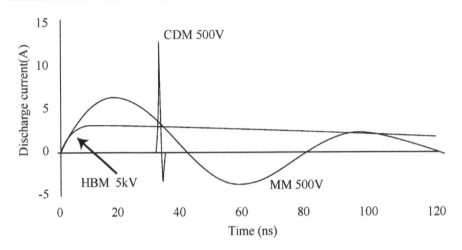

Figure 1.3. *Characteristics of HBM, MM and CDM discharge models*

	ESDA (ANSI) (ESD Association)	JEDEC (EIA) (Electronics Industries Association)	IEC (International Electrotechnical Commission)	AEC (Automotive Electronics Council)	EIAJ (Electronics Industries Association of Japan)
HBM	STM5.1-2001	JESD22-A114-E	61340-3-1 60749-26 ed2	Q100-002 REV-D	ED-4701/304
MM	STM5.2-1999	JESD22-A115-A	61340-3-2 60749-27	Q100-003 REV-D	
CDM	STM5.3.1-1999 STM5.3.2-2004	JESD22-C101-C		Q100-011	ED-4701/305

Table 1.2. *Overview of the various ESD standards at the component level (www.esda.org, www.iec.org, www.aecouncil.com)*

1.3. Standards at the system level

The number of standards used in the qualification of systems is quite low. Discharge models at the system level are aimed at reproducing ESD events that can take place during use or handling of an electronic system. This section presents the two standards and the associated testers, which are: IEC 61000-4-2 and ISO 10605.

It also provides information on the CDE (Cable Discharge Event) model, which is not normalized but still describes a real ESD event, and the HMM (Human Metal Model), which is being normalized and which describes a test method applied to the component.

As a summary we can say that, on the one hand, there are the component manufacturers, for whom ESD appears within the assembly areas. Integrated ESD protection methods are therefore adapted to these areas and are validated by the use of standards such as the HBM (Human Body Model), CDM (Charge Device Model) or the MM (Machine Model). The techniques used to develop protection methods are well controlled and rejections are relatively uncommon (Figure 1.4). Electronic circuits that are evaluated and tested during manufacturing are then placed onto electronic boards, which are in turn integrated into a complete electronic system such as a PDA, a computer, a mobile telephone, etc. However, the fact that a circuit survives a HBM/MM/CDM event during manufacturing does not mean it will survive the ESD events that the final product might be submitted to during everyday life [HYA 02].

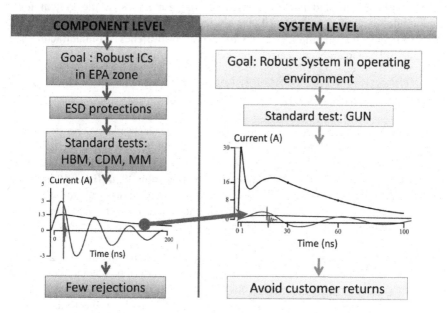

Figure 1.4. *Comparison of the issues of IC and EMs. For a color version of this figure, see www.iste.co.uk/bafleur/esd.zip*

On the other hand, system manufacturers must ensure the robustness of their applications in the environment of end usage. The main standard available is the IEC

61000-4-2 [IEC 08], whose energy and power level is much higher than the standards used for components. The right part of Figure 1.4 shows the system constraints and current levels that systems must withstand following the component standards.

Although this standard has been applied for 20 years, the problem of ESD in systems started being crucial around the years 2007–2008. The first issues really appeared around the year 2005, in automobile and military applications. In the period between 2005 and 2010, which corresponded with the advent of nomadic electronic systems, the impact of ESD at the system level became a major axis. To this day, the electric response of electronic systems to rapid transient electrical stress, also known as Electrical Fast Transient (EFT), of which ESD is a part, has not yet been characterized. At the same time, as a result of the development of semiconductor technology and the growth of embedded applications submitted to harsh environments, failures caused by ESD have become a major concern. ESD is a significant source of interference that can lead to electronic failure and loss of function, greatly reducing the reliability of equipment. The article published by J. Rivenc estimates around two discharges per day for an electronic system in an automobile, or around 5,000 to 10,000 discharges over the lifetime of the product [RIV 04]. This study marked the start of increased awareness in manufacturers of the issue of ESD in embedded electronic systems.

1.3.1. The IEC 61000-4-2 standard or ESD gun

The main standard used in the qualification of an electronic system or piece of equipment is the IEC 61000-4-2 standard, first created in 1984 under the name IEC 801.2, and renamed in 1995 as IEC 61000-4-2 [IEC]. The standard IEC 61000-4-2 is called the "Gun test". A final revision to this standard from 2008 [IEC 08] allows for specification of the maximal levels of electromagnetic emissions for electronic systems and is aimed at EMs. It therefore did not come from the ESD community and has a different approach from that used for characterizing components.

The aim is to reproduce the typical discharge of a human body through a metallic object located in the product under test. In order to reproduce this event, the standard specifies a waveform of the current that must be injected into the product being tested. It also defines how the material and setup required for the trials must be organized, the test and calibration procedures, as well as the failure levels. As a generalization, the equipment under test is placed onto an isolator and both are then placed onto a metallic plate. This plate is connected to a reference ground plane using two in series resistors of 470 kΩ.

This type of gun helps produce discharges by contact or in the air using a geometrical model that represents a human finger and which acts as an output electrode of the generator. Discharges in the air are the most realistic but the waveforms can vary depending on the distance between the circuit and the electrode and fluctuations in temperature, humidity or in the approach velocity onto the circuit pin [IWA 93]. These different parameters can have a significant effect on the pulse level measured during the ESD. The standard tied to this model IEC 61000-4-2 defines four levels of conformity (see Table 1.3), depending on the degree of robustness in regards to the voltage of the various inputs and outputs of the system. This table defines the levels of the two types of discharge:

– Air discharge: the trigger of the gun is pulled. It is then brought closer until it touches the surface of the product. A discharge into the air takes place during this procedure.

– Contact discharge: the electrode of the gun is placed onto the conducting part, which is a conductor to the outside. The trigger is then pulled and a discharge is injected directly into the product.

Level IEC 61000-4-2	Maximal voltage Contact discharge (kV)	Maximal voltage Air discharge (kV)
1	2	2
2	4	4
3	6	8
4	8	15

Table 1.3. *Conformity level for standard IEC 61000-4-2*

A photograph of an ESD gun is provided in Figure 1.5(a). It is made up of a battery, a control screen, a trigger, a grounding cable and a discharge tip representing the human finger. The ESD gun is a high-voltage generator that stores an electric charge in a 150 pF capacitor, which discharges into the system under test through a 330 Ω resistance, corresponding to the resistance of skin (Figure 1.5(b)).

The waveform here is very different from the one used in component tests, as shown in Figure 1.6. The standard shows a first current peak with a very rapid rise time (0.7–1 ns) that can reach up to several tens of amperes.

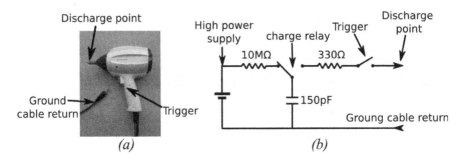

Figure 1.5. *Photograph of an ESD gun (a) and diagram of the tester (b)*

During calibration, the standard requires a target (Pellegrini target) whose equivalent resistance is 2 Ω. The waveform of the current produced by the gun must correspond to the reference waveform shown in Figure 1.6 and to the specifications provided in Table 1.4. Only three points on the time characteristic allow defining the discharge signal (first peak, amplitude from 30 to 60 ns). It can be noted that this waveform contains both characteristics of the HBM model in terms of its length and characteristics of the CDM model in terms of the very short rise time of the first peak.

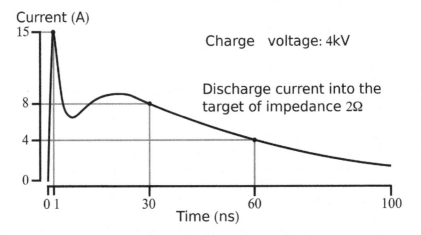

Figure 1.6. *Waveform of the discharge current of the ESD gun into an equivalent impedance of 2 Ω for a charge voltage of 4 kV*

Charge voltage (kV)	Current 1st peak at ±10% (A)	Current at 30 ns ±10% (A)	Current at 60 ns ±30% (A)
2	7.5	4	2
4	15	8	4
6	22.5	12	6
8	30	16	8

Table 1.4. Level of conformity for the standard IEC 61000-4-2

The normative document provides an equivalent electric diagram but it does not result in the previous waveform. The literature presents a large number of models, which are more or less precise and result in a more realistic aspect, such as the models by F. Caniggia [CAN 06], K. Wang [WAN 03, WAN 04a] or the model by K.M. Chui [CHU 03]. A comparative study of all of these models was carried out in the thesis by Sandra Giraldo [GIR 13]. This study shows that the model proposed by Chui is rather close to reality. It remains simple, flexible and contains few elements that are clearly associated with the parasitic elements of the earthing and of the gun, as reported in Figure 1.7. The parameters of the model are summarized in Table 1.5.

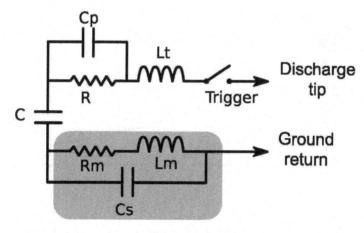

Figure 1.7. Equivalent diagram of the model of the gun [CHU 03]

In Figure 1.8, the stress IEC 61000-4-2 is compared to the standardized stress for components. The standards MM and CDM (500 V of charge in both cases – which is the level required for standard digital products) reveal a waveform with two alternating polarities but each with extremely different dynamics. The HBM standard corresponds to a long pulse with a rise time of roughly 10 ns spread over a hundred nanoseconds. The maximal current amplitude is 3.3 A (4 kV of charge) for the HBM. For the standard IEC 61000-4-2, the maximal current peak of 30 A is reached in less than 1 ns for a charge of 8 kV. From the perspective of the component standards (HBM and CDM), the robustness levels required by the IEC standard are far greater.

Elements used	Values	Comments
Gun capacitance (C)	150 pF	*Needed for the general structure*
Discharge resistance (R)	330 Ω	
Earthing inductance (Lm)	1.5 µH	*Impact on the 2nd peak*
Earthing resistance (Rm)	130 Ω	
Discharge tip (Lt)	75 nH	*Impact on the rise time*
Capacitance between gun and ground (Cs)	18 pF	*Impact on the waveform between the 2 peaks*
Internal parasitic capacitance (Cp)	3 pF	

Table 1.5. *Value of the parameters of the Chui model [CHU 03]*

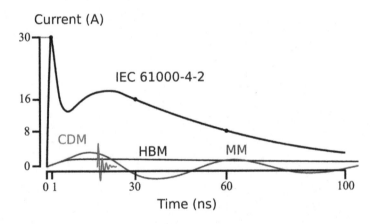

Figure 1.8. *Comparison of the dynamics of the discharge currents HBM (4 kV), CDM (500 V) and MM (500 V) and IEC 61000-4-2 (8 kV)*

Besides the injected powers, which are ten times more powerful, the qualification of the equipment must be done not only in a non-powered configuration (in line with the recommendations for the component) but also in a powered configuration. The functional robustness is defined by four classes A, B, C and D which define the reliability of the equipment:

– A: Normal operation within the limits specified by the manufacturer, the qualification applicant or the buyer.

– B: Temporary loss of functionality or temporary degradation of function stopping after the end of the disturbance; the equipment undergoing the test is then returned to its normal functions without intervention from an operator.

– C: Temporary loss of functionality or temporary degradation of function, whose correction requires intervention from an operator.

– D: Loss of functionality or degradation of function, non-recoverable, caused by damage to the hardware or software or by a loss of data.

The term "susceptibility" borrowed from the EMC field (ElectroMagnetic Compatibility) is also used in this work when the system is not destroyed by the ESD (Class A, B or C). When destruction takes place, the term used is "robustness".

1.3.2. Problems linked to the standard: IEC 61000-4-2

The standard only specifies three points of the dynamics of the current waveform. All the testers respect the standard and the measurements of these points are correct but variations outside of these three points are significant. It is therefore not surprising to see differences in robustness depending on the brand of the gun used [POM 95]. More in-depth studies analyzing the spectrum of the gun in conduction mode have shown variations of 10 dB depending on the frequencies. This issue of the reproducibility of IEC61000-4-2 tests has been shown by K. Muhonen on transmitted modules tested with two guns of different brands [MUH 09]. The total energy sent by these two testers is the same. However, depending on the ESD gun used the robustness of the product goes from 7 to 12 kV, which is due to the intrinsic radiation spectrum of ESD gun.

Several studies have also shown that ESD guns generate strong electric and magnetic fields [XIA 03, CHU 04, WAN 04a, WAN 03]. Measurements have been made by M. Honda to evaluate the transient magnetic and electric fields that can be coupled to a system during the application of a discharge with an ESD gun [HON 07]. The magnetic and electric fields measured are in the order of 9–90 A/m and 3–36 V/m, respectively. During the IEC61000-4-2 tests, these strong transient fields are coupled onto the tracks of the printed circuit board or to the connection cables of the product under test. This disturbance phenomenon is not linked to the direct injection of the stress onto one of the system inputs but to an electromagnetic interference type disturbance.

J. Koo, in collaboration with various researchers, carried out a comparative study in three laboratories, two located in the United States and one in Japan [KOO 08]. In the study, he analyzed the impact of the parameters of the discharge current as well as the transient fields of eight ESD guns on the IEC tests. The objective is to determine which parameters influence the reproducibility of the tests. Strong frequency variations can be seen from one gun to another. The various tests have led to the conclusion that the transient electromagnetic fields of the guns contribute significantly to the issue of the reproducibility of the tests at the system level.

1.3.3. HMM (Human Metal Model)

Increasingly, EMs are asking integrated device manufacturers (IDMs) to apply the system standard directly onto the component. These requirements are based on the hypothesis that if the components can resist a direct system discharge then this component will also resist up to the same level when introduced into a system. Several studies carried out by [THI 10] and [STA 09] attempt to link the standard IEC 61000-4-2 to the ESD component standards. The ESD results obtained at the level of the component, using the standards HBM and CDM, do not enable the establishment of a relation with the ESD system stress. The differences between the test procedures and the electric characteristics of the various waveforms mean that correlation between these different methods is unlikely.

The tests that comply with standard IEC 61000-4-2 have become widespread in directly testing components. Even if this standard is aimed at complete systems, EMs are increasingly asking IDMs to test components using this standard [MUH 09].

There are several reasons for this. On the one hand, the EMs want to get rid of external protection methods added to the printed circuit board in order to minimize production costs. On the other hand, in the automobile domain, circuits must be increasingly robust as a result of the increasing number of switching inductive charges (relays, motors, etc.) causing disturbances [CAO 10]. This is especially a requirement for the communication circuits CAN (Control Area Network) or LIN (Local Interconnect Network). These circuits, which are integrated in the many modules of a car, have one or several pins directly linked to an external connector. They are therefore susceptible to receiving very strong system type discharges.

Interpretation and implementation of system tests on the component inevitably vary from one manufacturer to another which is problematic in terms of interpreting the results. On top of this, there are the problems linked to ESD guns (especially radiation), described in the previous section, which also affect reproducibility. For the EMs and IDMs to be able to compare their data, normalization is needed. In order to comply with these requirements a new test method, called HMM for "Human Metal Model" and initialized in the article [CHU 04], is being discussed. A standardization committee (group SP5.6 of the ESD Association [ESD 09]) is currently working on its definition. Its objective is to define a test method to evaluate a component while using the waveform of IEC61000-4-2.

The emerging HMM standard [ESD 09] provides some information on the test procedures: a universal test board, the TFB (Test Fixture Board), has been developed (Figure 1.9). The electronic circuit is soldered onto the board and the impedance between the gun's injection points and the circuit is constant (50 Ω). One of the big differences compared to the IEC standard is that the earthing of the ESD gun is no longer connected to a coupling plane but rather directly to the ground of the printed circuit. This configuration eliminates the capacitance between the board and the ground, a capacitance that does not reflect reality [IND 10].

HMM tests are carried out following three different configurations:

– The tip of the gun is in contact with a point located on the TFB. The electromagnetic fields created by the ESD gun have an influence on the results.

– The TFB is fixed onto a metal plate, with holes at the areas corresponding to the discharge points behind the board. The tip of the gun passes through a hole and is placed in contact with a discharge point of the TFB. The metal plate acts as an electromagnetic shield and the fields radiated by the gun do not affect the measurements.

– The third configuration uses a 50 Ω generator instead of the ESD gun. The generator is connected to a test point using a 50 Ω cable. This enables total separation from the electromagnetic disturbances radiated by the gun and allows for total resolution of the issues of reflection linked to the impedance mismatching between the gun (330 Ω), the tracks of the board (variable around 100 or 50 Ω), and the chip (close to 1 Ω).

Tests have been carried out by the working group of the ESD Association in various laboratories around the world, looking at the reproducibility and validity of HMM tests. The results were published recently [SCH 16]. They show that the failures obtained in several components are similar from one tester to another.

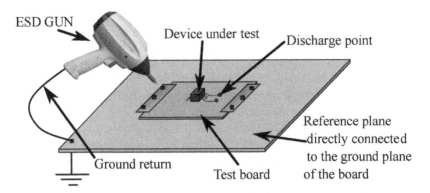

Figure 1.9. *Simplified diagram of the test setup used in the HMM [IND 10]. Courtesy of the Industry Council on ESD Target Levels*

1.3.4. *Standard model ISO 10605*

ISO10605 [ISO 08] is an automobile standard based on the standard IEC6100-4-2 described in section 1.3.1. It is an extension that specifies the test procedures aimed at integrated electronic modules in cars. Table 1.6 summarizes the different parameters used by this standard and the standard of IEC 61000-4-2.

Standard	ISO 10605	IEC 61000-4-2
Charge voltage (air mode)	2–25 kV	2–15 kV
Charge voltage (contact mode)	1–15 kV	2–8 kV
Time interval	1 s	1 s
Polarity	Positive and negative	Positive and negative
Capacitance of the ESD gun	150 pF/330 pF	150 pF
Resistance of the ESD gun	330 Ω/2 KΩ	330 Ω
Minimum number of discharges	3	10
Reference of the ESD gun	Vehicle battery	Ground
Test condition	Powered/Non-powered	Powered

Table 1.6. *Comparison of the test parameters of standards ISO10605 and IEC 61000-4-2 [IND 10]*

Firstly, the injection levels are clearly higher than those for the IEC standard. The discharge takes place through several of the RC modules, where all of the combinations between the capacitances of 150 Farad (F) and 330 pF and the resistances of 330 Ω and 2 kΩ are possible. This considerably changes the dynamics of the signals, as shown in Figure 1.10, and thus increases the harshness of the tests for rapid signals (≤ 1 ns) with high amplitudes and also for longer signals with smaller amplitudes (well over 100 ns).

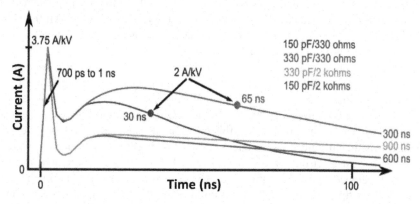

Figure 1.10. *Comparison of the waveforms imposed by the standard ISO 10605. The times given at the end of the curves correspond to the respective durations of the pulses. For a color version of this figure, see www.iste.co.uk/bafleur/esd.zip*

The tests are carried out under two conditions: non-powered and powered with a battery. The standard specifies how to connect the ground of the ESD generator to a coupling plane acting as the mass of the battery or to the chassis of the vehicle, unlike IEC61000-4-2 which specifies a connection to the ground.

1.3.5. CDE (Cable Discharge Event) model

Nowadays, electronic interfaces are vital for communication between systems. USB, FireWire, Ethernet, etc. are regularly used in nearly all electronic applications. With the rise in cabled networks, the network equipment or the Ethernet interface is becoming increasingly important [PIS 05]. Since the wiring of the Ethernet network is often very long, and contains twisted cables, the electronics involved in the Ethernet interface or modem is likely to undergo an ESD event called Cable Discharge Event (CDE). CDE is a discharge that takes place when a charged cable is connected to a part of an electronic system, such as the connection of the Ethernet cable to the interface [LAI 06] or of USB cables to an electronic device [STA 07]. The cable is susceptible to charging through a triboelectric process as it is handled during its installation in a building. The waveform associated with the ESD stress is very different from HBM, CDM or MM.

Unlike discharges that consist of a model with low capacitances and high impedances, the CDE discharges a very large quantity of charges with low source impedance. The static voltage that appears in the cable is determined by the triboelectric charging that is produced and the capacitance created between the cable and the external environment. However, there is very little information published on the waveforms of the discharge generated in or on the cable [BRO 01, SMI 02, CHA 06, STA 07]. These publications mainly present experimental results concerning the impact of an electronic system on the discharging of a cable of fixed length and charged at a given voltage. However, it has been shown that CDE discharge starts with a current spike followed by an oscillatory phenomenon, whose frequency is dependent on the length of the cable: the shorter the cable, the quicker the oscillations. A CDE stress can lead to logical faults or failures in equipment or electronic systems. The issue with this phenomenon is that there are no test standards for quantifying the characteristics of this discharge. In order to evaluate the robustness of electronic systems with regard to CDE, a measurement method must be developed that can capture the events, record them and analyze the data obtained. With this aim, the working group "IEEE 802.3 Cable Discharge Ad-hoc" was created in January 2001 to respond to demands from manufacturers regarding failure rates of LAN (Local Area Network) devices and products undergoing CDE [WOR 01]. To this day, there is no normalized model and method due to the multiplicity of the case studies.

1.4. Conclusion

Guaranteeing a certain level of ESD robustness in an electronic system involves, on the one hand, ESD qualification of its different components and, on the other hand, ESD qualification of the complete electronic system. In order to do this, a certain number of standards have been developed.

Component level standards HBM, MM and CDM ensure that the components, which are the ICs here, are able to survive all steps of manufacturing and assembly up to mounting on an electronic board. These tests are always carried out on components that are not powered.

ESD qualification at the system level must allow for evaluation of the robustness of the system with regard to stresses issued in the application environment which can sometimes be very harsh, as in automotive settings, for example. The relevant tests are carried out in powered and non-powered configurations.

Equipment manufacturers first required a level of ESD robustness for elementary components in contact with the external connectors close to the levels required for the system. It quickly became apparent that there is little or no correlation between the HBM robustness of a component on a board and its ESD robustness.

These ESD qualification tests allow us to evaluate the level of ESD robustness but do not help us understand why some components pass the test and others do not. In order to better understand the origin of ESD failures and to apply the necessary corrective measures, specific characterization tools have been developed, as well as new techniques to locate generated defects. All of these techniques are presented in the following chapter.

2

Characterization Techniques

ESD qualification tests do not provide information regarding the origin of a failure and therefore do not help set up corrective measures to improve the component or system ESD robustness. In this chapter, we shall review the current state of characterization, injection and measurement techniques that have been developed globally and at the LAAS-CNRS, on the one hand, to characterize ESD protection structures, and on the other to characterize the response of a circuit or of a system to ESD stress.

One of the main techniques is the use of a TLP (Transmission Line Pulsing) test bench, which helps generate pulses controlled by amplitude, duration and rise time. Although the base principle remains the same, there are several possible variations and uses for TLP, which are presented here. These techniques are inspired from basic device measurement principles such as TDR (Time Domain Reflectometry), modified or adapted for use in ESD. These techniques are essential, first for retrieving information on what is going on in the component or system tested during discharge, but also to check and validate the models developed in Chapter 3. Other injection and measurement techniques inspired from ElectroMagnetic Compatibility (EMC) standards are also presented. These standards have been adapted for the characterization of the behavior of systems undergoing ESD stress.

We shall also describe a number of techniques used in the localization of induced failures, notably those based on laser stimulation. Finally, we shall look at the potential of low frequency noise characterization for detecting latent defects. These techniques are essential tools for in-depth analysis of the behavior of a protective structure during ESD stress, extraction of the main parameters (R_{ON}, trigger voltage, trigger dynamics, etc.) and validation of the global ESD protection strategy.

2.1. Component level electrical characterization techniques

The robustness level of a component in terms of ESD is usually evaluated by carrying out a characterization with test equipment that follows the standards [ANS 10, ANS 99a, ANS 99b]. For the HBM and MM standards, the circuits, assembled in their packaging, receive a calibrated waveform through a complex connection matrix that allows for the various pin combinations to be tested. Currently, the high number of inputs–outputs is a new source of issues for the test protocol [STA 15]. Testing standard CDM requires specific equipment to enable the global charging of the component and a robotic arm to sequentially trigger the discharge into each of its pins.

These test benches help prove the efficiency of a protection strategy for a circuit, but do not provide the information required for development and optimization of the protection structures, such as the trigger voltage, resistance during the "on" state or even trigger dynamics. Extraction of these intrinsic parameters is not without difficulty. Direct measurements from the test bench are made impossible by the presence of many noise sources and by the difficulty of placing the test probe as close as possible to the component. Ideally, the measurement is made under the tip. However, traditional static I-V measurement techniques cannot be used be as the thermal heating of the component due to the high current levels involved during ESD would lead to its premature destruction. Moreover, the trigger dynamics would be completely hidden. The static I-V measurement is only used for weak currents to show the degradation of the tested component. Characterization of the components is therefore done through analysis of the response to a current pulse.

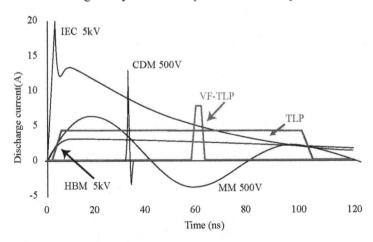

Figure 2.1. *Comparison between ESD stress and TLP/VF-TLP pulses. For a color version of this figure, see www.iste.co.uk/bafleur/esd.zip*

2.1.1. *TLP/VF-TLP measurements*

TLP (Transmission Line Pulse tester) generates voltage pulses that last 100 ns and can reach currents of several amperes in the studied structure if triggered. The rise time can be controlled using a passive filter to simulate a HBM discharge as much as possible. The principle relies on the theory of transmission lines, applied to ESD characterization [MAL 85]. The advantage of using a transmission line compared to a "solid state" pulse generator is two-fold: relative ease of obtaining a large range of current levels, and a relatively large system tolerance to variations of the charge impedance. The test component does initially present a very high impedance level, which is then greatly reduced following the triggering. Pulse duration, however, is fixed by the length of the line. The VF-TLP uses the same principle as the TLP, but it is adapted to shorter pulse durations. Figure 2.1 provides a comparison of the waveforms of these pulses compared to the ESD standards. First developed in laboratories with no shared reference, the test benches are not normalized but now do share reference documentation [ANS 07, ANS 14b].

Measurement of the voltage and of the current in the component being tested is usually done using the TDR method. The calculation is carried out using the measurement of the incident pulse of the generator and the pulse reflected by the component (Figure 2.2). The voltage is measured directly using a high-bandwidth oscilloscope in single-shot mode through an attenuator. The current is measured in another channel of the same oscilloscope using a current probe.

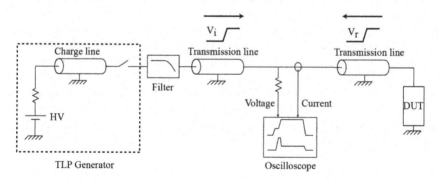

Figure 2.2. *Principle of TDR measurement for the TLP*

In the case of TLP, the connecting cable between the generator and the component is short in relation to the duration of the pulse (100 ns). The incident and the reflected waveforms are therefore superimposed, and the measurement can be carried out directly using the oscilloscope's trace. For a given charge voltage in the coaxial line, an I-V couple is extracted at the end of the pulse. By changing the values of the charge voltage, a complete quasi-static I-V curve can be constructed (Figure 2.3).

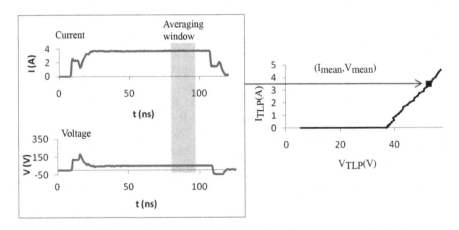

Figure 2.3. *Reconstruction of the quasi-static I-V curve from the TLP pulses. For a color version of this figure, see www.iste.co.uk/bafleur/esd.zip*

In the case of the VF-TLP, measurement is a bit more complicated because, as the pulse is short (1–5 ns), the incident pulse and the reflected pulse are separated. Reconstruction therefore takes place either mathematically during post-processing, or analogically using a delay line. In any case, the system must be well-calibrated, temporally, by not only precisely determining the delays induced by the coaxial cables and the measurement points, but also by characterizing the losses in the system and the values of the attenuators used. The VF-TLP is mainly dedicated to measurements of the component at the wafer level. In order to obtain good results, radiofrequency probes are used, thus getting rid of parasitic elements and providing clean characteristics. However, their use requires the implementation of specific pads on the chip.

There are more complex configurations that improve the performance of the VF-TLP, such as the TDR-T. An additional channel of the scope is used for the ground of the structure being tested, as shown in Figure 2.4. This transforms the impedance of the system to 100 Ω and provides both the advantages of direct measurement of the transmitted current as well as those of a differential measurement at the component terminals.

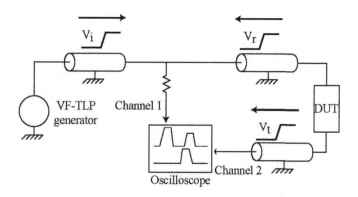

Figure 2.4. *VF-TLP configuration in TDR-T*

Some authors have put forward correlations between the failure level I_{T2} of the TLP and the HBM robustness, but others refute this [STA 98, GUI 05]. However, there is no clear correlation between the VF-TLP and the CDM [GIE 98]. Studies have been carried out on a new measurement method, the CC-TLP (Capacitively Coupled TLP) [WOL 05], at chip level. The pulses are directly injected in the pad to be tested, while the component is connected to the ground via a packaging emulator. High levels of correlation are found with the CDM on the packaged circuits, as much for the peak current value as for signatures of failure. However, this system, which is hard to put into practice, did not go beyond the laboratory test bench. The VF-TLP remains a good method for testing the triggering levels and intrinsic robustness levels of an ESD protection system.

2.1.2. The transient-TLP

The different TLP systems presented previously operate in a quasi-static state. Each couple (I, V) is extracted near the end of a pulse, where the signal is usually stabilized. If the structure is properly designed it should not present any overvoltage in relation to its trigger point, thus guaranteeing the validity of the TLP curve. However, for simple protection systems, notably in high-voltage applications, these overvoltages do exist, and it is worth characterizing them. A specific set-up, derived from the VF-TLP, was developed to measure these overvoltage peaks [DEL 12]. The test set-up is reduced to the minimum (Figure 2.5): a high-voltage (HV) generator and its line TL_1 that generate pulses of 5 ns, a pick-off T attenuator to extract the voltage and take it through a transmission line TL_3 toward a channel of the oscilloscope and the transmission line TL_2 to connect to the component, or DUT (Device Under Test).

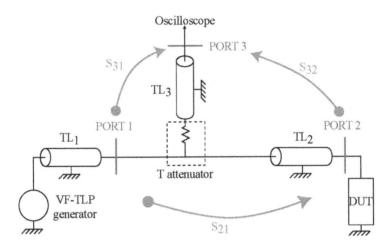

Figure 2.5. *Transient-TLP test set-up and its transmission parameters S*

First, the system is characterized by a vectorial network analyzer, in order to extract the transmission parameters S_{21}, S_{31} and S_{32}. Next, the incident and reflected pulses measured by the oscilloscope are taken into the frequency domain using a Fourier transform, respectively becoming $V_{inc,oscillo}(f)$ and $V_{ref,oscillo}(f)$. The voltage at the DUT is then calculated as:

$$V_{DUT}(f) = \frac{S_{21}}{S_{31}}.V_{inc,oscillo}(f) + \frac{1}{S_{32}}.V_{ref,oscillo}(f) \tag{2.1}$$

The current is obtained by considering the characteristic impedance Z_C of the coaxial cables:

$$I_{DUT}(f) = \frac{1}{Z_C}\left(\frac{S_{21}}{S_{31}}.V_{inc,oscillo}(f) - \frac{1}{S_{32}}.V_{ref,oscillo}(f)\right) \tag{2.2}$$

An inverse Fourier transform is applied to move back into the time domain. Figure 2.6 illustrates the performance of the test bench for three rise times of the generator (V_{charge} = 250 V, duration = 5 ns) by presenting the voltage (a) and the current (b) within a protection system [DEL 10]. The overvoltage can be seen to be dependent on the rise time, and presents a maximal value of 200–300 % of the quasi-static value that is extracted by a VF-TLP.

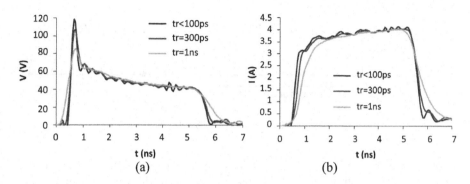

Figure 2.6. *Measurement of an ESD protection system on the Transient-TLP [DEL 10]. For a color version of this figure, see www.iste.co.uk/bafleur/esd.zip*

It is more likely that these excessive values, which are non-destructive for the protection system, greatly reduce reliability of the protected circuit in the long term.

2.1.3. Adaptation of the TLP for reflectometry

2.1.3.1. Principle of measurement by reflectometry

The principle of reflectometry involves inserting a probe (voltage or current) and then observing the reflections in the time domain. This technique can be compared to the use of radar, where after having transmitted a pulse, the echo that is returned (reflected signal) provides the distance of an object, such as a boat or an airplane, as well as its characteristics. In our case, the reflection provides information on the impedances encountered and their distance in relation to the source of the emissions [AGI 00]. Initially, these analyses would determine the length of a uniform cable or detect defects located in the cable by sending a pulse and measuring the attenuation of the pulse sent back, along with the response time.

Before going further ahead in the use of the method of TDR as applied to ESD, first we must define the notions of propagation and reflection coefficient. To do this, we shall look at the case of a transmission line. An elementary section of a transmission line is usually represented with the elements R, L, C and G. These are discretized in n portions so as to model a line along the TEM (Transverse-ElectroMagnetic) approximation [IEC 05], as shown in Figure 2.7.

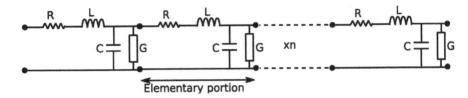

Figure 2.7. *Classic RLCG model of a transmission line*

If we consider a portion of an infinite line, contained between x and δx, using the Maxwell equations, the equations for the propagation of the voltage V and of the current I for a transmitted sinusoidal signal are written [CHA 13, PER 96, CLA 92] as:

$$\frac{\partial^2 V}{\partial x^2} = \gamma^2 V \text{ and } \frac{\partial^2 I}{\partial x^2} = \gamma^2 I \qquad [2.3]$$

$$\text{with} \quad \gamma^2 = (R + jL\omega)(G + jC\omega) \qquad [2.4]$$

where γ is the propagation constant. This constant is complex and can be put into the following form:

$$\gamma = \alpha + \beta j = \sqrt{(R + jL\omega)(G + jC\omega)} \qquad [2.5]$$

The voltage introduced by the source at x requires a finite amount of time to go through the line to δx. The phase of the voltage at point δx is shifted by the factor β in relation to the source voltage in x. The phase shift β is expressed in radians per meter. Moreover, the voltage is attenuated by a factor α expressed in dB per meter.

In the case of lossless lines, R=0 and G=0, the attenuation coefficient α is equal to zero. In this case, the propagation constant is worth $\gamma = j\omega\sqrt{LC}$ with the phase shift $\beta = \omega\sqrt{LC}$.

The phase or propagation velocity is written as:

$$V_p = \frac{\omega}{\beta} \text{ in m/s} \qquad [2.6]$$

or, for the lossless lines [BAK 90]:

$$V_p = \frac{1}{\sqrt{LC}} \text{ in m/s} \qquad [2.7]$$

The propagation constant is used to define the current and the voltage for any distance x from the source, with the following relations (general solutions to the equations of [2.3]):

$$V_x = V_{incident} * e^{-\gamma x} \text{ and } I_x = I_{incident} * e^{-\gamma x} \qquad [2.8]$$

With $V_{incident}$ and $I_{incident}$ as the incident voltages and currents, respectively. The voltage and the current are tied to any distance x by the characteristic impedance of the line:

$$Z_x = \frac{V_x}{I_x} = \frac{V_{incident}}{I_{incident}} \qquad [2.9]$$

In the case of a lossless line (negligible resistive and dielectric losses), the characteristic impedance of a line can be simplified:

$$Z = \sqrt{L/C} \qquad [2.10]$$

As the board has a continuous ground plane, the tracks are considered to be microstrip lines with a characteristic impedance of Zc as soon as the signals that they carry are of high frequency. The definition of what is a "high frequency" still must be set and is relative. The microstrip line is discretized into n elementary portions as illustrated in Figure 2.7. This discretization must take into account the quasi-TEM approximation [IEC 05]. As a first approximation, the mode of propagation of the wave along the line is quasi-transverse, meaning that the electric and magnetic fields are perpendicular to the axis of the line. In this approximation, the phase variations are neglected. Thus, the length l of the elementary element δx must fulfill the following condition:

$$l \ll \frac{\lambda}{4} \text{ with } \lambda = \frac{v}{f} \text{ and } v = \frac{c}{\sqrt{\varepsilon r}} \qquad [2.11]$$

where λ is the wavelength, f is the frequency, v is the velocity and finally c is the speed of light (c=3×10^8 m/s).

By considering the work frequency to be 2 GHz, the wavelength is 7 cm. The elementary length δx must be very much lower than λ/4 ≈17.5 mm. To obtain a sufficient condition, we take dl = λ/10 ≈ 7 mm [LAC 08].

With ESD phenomena whose rise time is in the order of the nanosecond or less, it is important to consider the propagation delay, which is not negligible for several centimeters of tracks on a printed circuit board. These delays can result in voltage peaks in the measurements depending on the reflection [BÈG 14a, BÈG 14b].

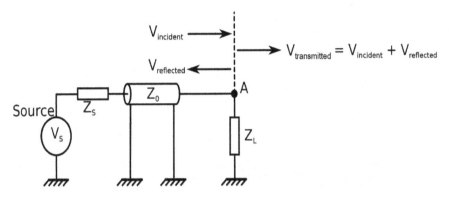

Figure 2.8. *Example of impedance loss: impedance transmission line Z_0 ending with an impedance of a different value Z_L*

When a transmission line of finite length and of impedance Z_0 is terminated by a charge Z_L (Figure 2.8), two cases can arise:

– The charge is adapted, meaning that characteristic impedance Z_L is the same as Z_0. Equation [2.7] is verified.

– The charge is not adapted; the impedance Z_L is different from Z_0. Part of the incident wave ($V_{incident}$) that arrives at point A is reflected and propagates from the charge toward the source ($V_{reflected}$). The observed impedance at break point A is then worth:

$$Z_L = Z_0 \frac{V_{reflected} + V_{reflected}}{V_{incident} - V_{reflected}}$$ [2.12]

The transmitted voltage ($V_{transmitted}$) is worth:

$$V_{transmitted} = V_{incident} + V_{reflected}$$ [2.13]

The ratio between the reflected wave and the incident wave is called the reflection coefficient Γ and depends on the characteristic impedance of the line Z_0 and of the charge Z_L following expression [BAK 90]:

$$\Gamma = \frac{V_{reflected}}{V_{incident}} = \frac{Z_L - Z_0}{Z_L + Z_0} \qquad\qquad [2.14]$$

TDR is a measurement technique used to analyze impedance and locate defects in transmission lines, connectors, printed circuits and other electric pathways [DAS 96]. The principle involves sending a voltage or current wave, and to observe the reflections in time domain. The diagram describing the principle behind TDR is provided in Figure 2.9.

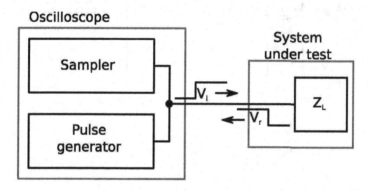

Figure 2.9. *TDR device principle diagram*

Reflection is a phenomenon that will likely modify the shape of a signal propagating along a line. The incident wave V_i is propagated following the propagating velocity of the line (equation [2.7]) toward the charge Z_L. Any discontinuity in the propagation path of the line will cause the system under test to reflect part of the incident wave according to the reflection coefficient (equation [2.14]). The reflected wave V_r is propagated toward the source.

An example of a TDR study is provided in Figure 2.10. A source V_S of impedance Z_S is connected to the end of a transmission line of impedance Z_0 whose propagation time is $T_\rho = 1$ ns. The charge is $Z_L = 1/2 \times Z_0$. At the point A, the loss of impedance leads to reflections whose coefficient is:

$$\Gamma = \frac{(\frac{1}{2}Z_0) - Z_0}{(\frac{1}{2}Z_0) + Z_0} = -\frac{1}{3} \tag{2.15}$$

The reflection that appears at the time t_1 is $V_{reflected} = \Gamma \times V_{incident} = -1/3 \times V_{incident}$. The time between t_0 and t_1 corresponds to the time taken by the incident wave to get onto the charge and to be reflected toward the source, so two times the propagation time of the line.

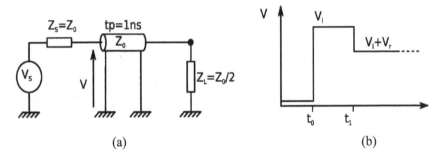

(a)   (b)

Figure 2.10. *Example of a TDR analysis: corresponding diagram (a) and chronogram (b)*

The limitations of this TDR method involve the multiple effects of reflections that appear in circuits that contain several interconnections or impedance changes. At each impedance discontinuity, part of the incident signal propagating through the system under test is reflected, and as a result, only the other part of this signal is transmitted onto the following discontinuity. Several impedance changes lead to multiple reflections (Figure 2.11) [SMO 99]. At each loss (moving from Z_i to Z_{i+1}), part of the incident wave is reflected and goes back to the TDR measurement instrument. As soon as it encounters another loss, it reflects again toward the system under test.

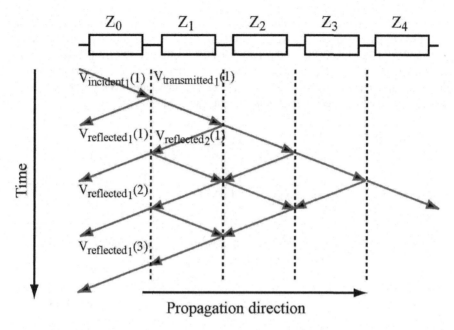

Figure 2.11. *Multiple reflections on a track containing discontinuities*

2.1.3.2. *Application of the TDR method to ESD events*

TDR can help obtain information more rapidly on the behavior of a system undergoing an ESD event. The biggest advantage of this method is that it only involves the elements that are external to the system under test (generators, probes and lines), and there is no need for specific patterns to be integrated in the printed circuits. In order to obtain a simple form to analyze, we used a TLP bench to create a rectangular pulse. The gun was not used in this measurement method as its waveform is complex and difficult to analyze. A diagram describing the principle of the TDR test bench using the TLP is presented in Figure 2.2.

The TDR/TLP bench uses exactly the same principle as the TDR measurement bench, with the only difference being the latter is configured to send only one pulse. The quasi-static aspect is not used in this very powerful operating mode, thus representing the behavior of the system undergoing an ESD event. We only look at waveforms that are dynamic, incident (V_i, I_i) and reflected (V_r, I_r), measured using probes inserted into the generator and the system under test.

The very high intensity values of the current in the discharges, as well as the large spectrum induced by the ESDs require the use and/or development of specific probes. The current probes used are the CT1, CT2 or CT6 probes made by Tektronix

[SPÉ 06]. The probes are current loops. A small hole allows the conductor being measured to pass through. The voltage induced in the loop is directly proportional to the current. The sensitivity of the three probes useful for ESD measurements are summarized in Table 2.1, as well as their main characteristics. Depending on the probe used, current peaks of 6–36 amperes can be measured. Measurement of very strong currents, however, is detrimental to the bandwidth.

Characteristics	CT1	CT2	CT6
Bandwidth	25 KHz to 1 GHz	1.2 KHz to 200 MHz	250 KHz to 2 GHz
Rise time	350 ps	500 ps	200 ps
Sensitivity	5 mV/mA	1 mV/mA	5 mV/mA
Max current peak	12 A	36 A	6 A

Table 2.1. *Main characteristics of the appropriate probes for measuring strong ESD currents*

The test bench is adapted to 50 Ω and the measurement system (TDS 6604B oscilloscope, RG 402/U cables) has a bandwidth of 6 GHz. In this test bench, we use the VF-TLP as the pulse generator. The voltage and current probes are integrated into a measurement module that connects the bench to the oscilloscope. Very short pulses (from 1.25 to 5 ns) with very short rise times (in the order 100 ps) are injected onto the lines through SMA connectors.

A complete model of the TDR/TLP measurement set-up was developed in order to compare measurements and simulation, and thus to obtain the relative error of our line models (Figure 2.12).

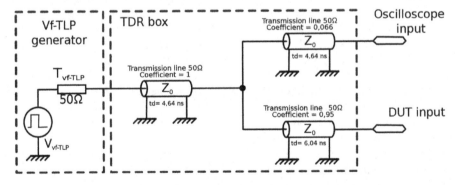

Figure 2.12. *Electrical model of the injection bench: VF-TLP with delays and attenuations [LAC 08]*

In Figure 2.12, the VF-TLP bench is modeled using a pulse generator followed by a resistor of 50 Ω. The model of the cables is designed using a transmission line with a propagation velocity of 1.94×10^8 m/s, which gives a delay of around 5 ns per meter. For the model of the measurement module, various characterizations in [S] parameters were carried out in order to see the existing attenuations [LAC 08]. The measurements show a slight attenuation in the pulse transmitted between the module input and the output linked to the board. An attenuation of −35 dB can be seen in the reflected pulse sent to the oscilloscope. A propagation difference of around 1.4 ns is present between the pulse sent to the oscilloscope and the pulse transmitted to the board.

By using coaxial cables with propagation times that are greater than the width of the emitted pulse, the incident wave is not superimposed onto the various reflections caused by mismatches of the line. This makes the interpretation of the discontinuities along the line (bends, vias, crossings) easier. As a result, for this type of characterization the use of the VF-TLP bench is well-suited because of the injection of short and rapid pulses (duration of 5 ns, rise time of ≤100 ps).

2.1.3.3. Application of the TDR/TLP tool

The technique of TDR reflectometry tied to a pulse generator like the VF-TLP (Very Fast TLP) bench [GIE 98, GRU 04, LAC 07a] is a very interesting tool for validating line models in the operating state of ESDs (high power, high frequency).

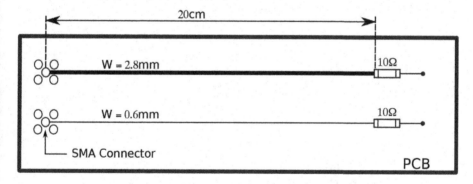

Figure 2.13. *Example of a printed circuit board used in the study of line propagation models*

Several line patterns were studied in the thesis by N. Lacrampe with the goal of validating the propagation models of the tracks of printed circuit boards [LAC 08]. A representative example of the studies carried out is shown in Figure 2.13. It represents two lines with different characteristic impedances, loaded by a resistor of equal value. Other line shapes with different characteristic impedances are analyzed, with or without discontinuities. Some patterns also help study not only the unique cases of vias, bends and the presence or absence of ground planes, but also diaphonic couplings between tracks. In order to validate the models (Figure 2.14) that we shall use hereafter in predictive system simulations, comparisons between simulations and measurements (Figure 2.15) were carried out.

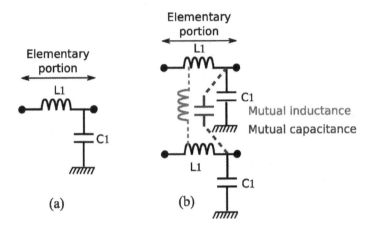

Figure 2.14. *Elementary models used for modeling single line (a) or coupled lines (b). For a color version of this figure, see www.iste.co.uk/bafleur/esd.zip*

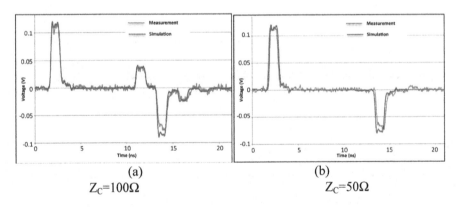

Figure 2.15. *Measurement and simulation of a line of length 20 cm, of characteristic impedances 100 Ω (a) and 50 Ω (b) with a resistive load of 10 Ω. For a color version of this figure, see www.iste.co.uk/bafleur/esd.zip*

In Figure 2.15, the first pulse observed corresponds to the incident signal. The second pulse is the reflection. For a line impedance of 100 Ω, it is first of all positive, corresponding to the line impedance, and then negative, which represents the resistive load of 10 Ω. For a line with a characteristic impedance of 50 Ω, only the reflection in the terminating resistance is apparent. The approximation of LC used for the propagation of the ESDs correlates perfectly with the measurements for the propagation along the lines. A small error is observed for the second reflected pulse. This is most likely due to high frequency parasitic elements of the soldered resistor at the end of the line, notably the parasitic series inductance.

Line models have been developed to take into account coupling or diaphonic phenomena between two tracks that are close [FLE 88, BAK 90]. The theory of coupling between two tracks, which aids in the development of simulation models, is provided in section 2.2.3.1. The different types of coupling (capacitive and inductive) depend on the configuration of the present conductors, i.e. the geometric characteristics (distance between the tracks, orientation, presence or absence of a ground plane). As for the previous model, a generic model called "LC_coupled" (Figure 2.14) is implemented, whose input parameters are the linear as well as mutual capacitances and inductances.

Figure 2.16 shows the comparison between the measurement and simulation of a line coupled with itself. The simulation is well in line with the measurement when the implemented LC elements come from analytical formulae that are appropriate for the calculation of these elements in configurations corresponding to the propagation of slow waves [BAK 90]. The results provided by these formulae are just as appropriate for studying the propagation of ESDs, as the frequency range used remains small (less a few GHz). The simulations do not show phenomena of dielectric losses or skin effects.

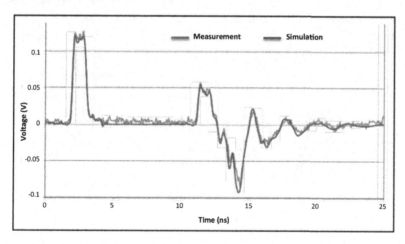

Figure 2.16. *Measurement and simulation of a line coupled with itself. For a color version of this figure, see www.iste.co.uk/bafleur/esd.zip*

2.2. System measurement methods

In order to characterize the impact of ESDs at the level of electronic systems (i.e. an electronic board), current and voltage measurement tools are needed to analyze propagation phenomena within the system. Three techniques are provided here, whose goal is to characterize the currents and voltages at the terminals of components. Two of these require the development of specific boards for inserting measurement elements, such as resistors or coupled lines. Only the near-field measurement technique can be used on real products.

2.2.1. Voltage measurement by probes

A probe in a printed circuit, the drawing of which is given in Figure 2.17(a) can be used for measuring the voltage. The probe is made up of tracks present on the printed circuit board of impedance $Z_0 = 50\ \Omega$ and of a resistor of R = 450 Ω. The substrate is FR4 epoxy, with a thickness of 0.8 mm. The connectors are SMA. The voltage measured at port 2 corresponds to the voltage that runs along the line between port 1 and 3, attenuated by 20 dB. The attenuation α is set using the following relation:

$$\alpha = \frac{R + Z_0}{Z_0} = \frac{450 + 50}{50} = 10 \qquad [2.16]$$

In practice, we have chosen a "high frequency" resistor with a normalized value of 470 Ω. The resulting attenuation is of approximately 20.3 dB. A measurement of the frequency response is given in Figure 2.17(b). This measurement, provided by the spectrum analyzer, allows checking the attenuation coefficient as well as the linearity of the frequency. According to the measurement, the linearity is very good up to 5 GHz, which is sufficient for ESD measurements.

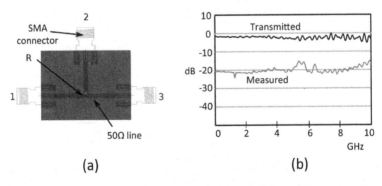

(a) (b)

Figure 2.17. Diagram (a) and frequency characteristics of the probe when powered (b). For a color version of this figure, see www.iste.co.uk/bafleur/esd.zip

2.2.2. Measurement of grounding currents through a 1 Ω method

In order to measure the current circulating throughout a circuit, the measurement commonly called the "1 Ω measurement", inspired from part 4 of standard IEC 61967 [IEC 07a], can be used. This standard initially defined a determination method for the conducted electromagnetic emissions of an integrated circuit. This method has since been adapted to measure electrostatic discharge. The principle of the measurement consists of adding a resistance of 1 Ω on the path between the pin V_{SS} of a circuit under test and the ground plane, as illustrated in Figure 2.18. The current coming from the circuit and traveling toward the ground results in a potential difference between the terminals of the resistor (voltage V). A probe formed by the 1 Ω resistor and a second resistor of 49 Ω, thus ensuring 50 Ω matching, allows this voltage to be measured. The impedance ratio of the 50 Ω cable linking the measurement at the oscilloscope to the impedance of the probe divides the measured voltage by a factor of two.

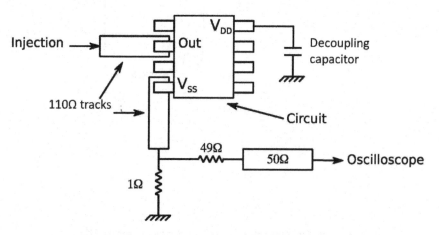

Figure 2.18. *Schematic representation of the principle behind "1 Ω" measurements [IEC 07a]*

An example of a current measurement using this method is provided in Figure 2.19. It is in fact an application of the diagram of the principle shown in Figure 2.18. A TLP pulse of 1 A, 100 ns, with a rise time of 1 ns is injected into the circuit output. Measurement of the current is carried out in two cases: with and without a decoupling capacitor of 50 nF connected between the pin V_{DD} and the ground plane. Without the decoupling capacitor, all of the current injected into the output is evacuated through the pin V_{SS} (Figure 2.19). With the decoupling capacitor, however, the current is spread out between the capacitor and the pin V_{SS}. The dynamics of the current evacuated through the pin V_{SS} become complex

(Figure 2.19). This waveform, as well as all of the interactions that appear between the capacitor and the circuit have been interpreted, and are presented in detail in section 5.3 of Chapter 5 (Case no. 3: impact of decoupling capacitors in propagation paths in a circuit). As we shall see in that section, this measurement method allows for rapid transient phenomena induced by the component package to be observed. The bandwidth of this method is in the order of 1 GHz. Beyond this frequency, the impedance of these resistors varies.

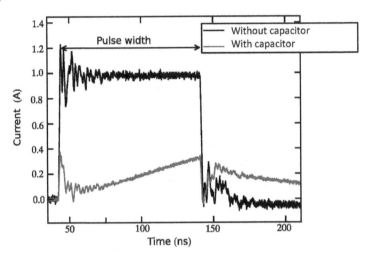

Figure 2.19. *Measurement of the current traveling through the circuit using the 1 Ω measurement, comparison with and without decoupling capacitor*

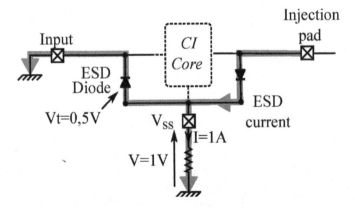

Figure 2.20. *Electrical diagram showing the intrusive aspect of the 1 Ω measurement*

This measurement method should be used with precaution as it is intrusive and shifts the reference potential of the circuit. This shift can result in a conduction path for the ESD current as part of the protection strategy of the circuit that does not represent reality. For example, a protective diode located between the rail V_{SS} and an input can be set off, as illustrated in Figure 2.20. The input is linked to the ground. If the threshold voltage of the ESD protection diode is 0.5 V, a current of 1 ampere across the 1 Ω resistor would result in a voltage of 1 V in the pin V_{SS}. The diode is then set off and the current is also evacuated toward the ground via IN pins, following a less resistive path.

In order to limit the voltage drop onto the pin V_{SS}, and considering the strong currents involved during discharges, a resistor of 0.1 Ω can be used instead of the 1 Ω resistor. The resistor of 49 Ω is then in turn replaced by one of 49.9 Ω in order to maintain an impedance matching of 50 Ω.

Another solution involves replacing the 1 Ω resistor by a near field probe. This solution is described in the following section.

2.2.3. *Measurement of currents by induced magnetic field*

In order to measure the strong ESD current going through the tracks, a probe that uses diaphony between two close tracks is actually a good alternative to the 1 Ω method. Our preferred form of coupling is inductive coupling as it is directly linked to the current running along the metallic tracks. However, reconstituting currents based on field measurements require the use of mathematical algorithms and precise sampling. This section is split into three parts. First of all, we must recall the theory on coupling between two close tracks. Two current reconstitution techniques are then shown, one temporal and the second based on frequency. Finally, we shall describe the validation of the method carried out, using measurements conducted along a sampling pattern.

2.2.3.1. *Magnetic coupling – theory*

Current and voltage transitions along a disturbing track generate magnetic and electrical fields that couple and induce disturbances along a victim line. When two tracks are close, a mutual influence is exerted between them mainly by:

– Electric field creating a capacitive effect between the tracks;

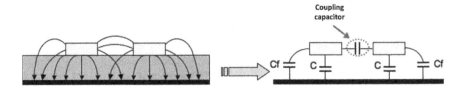

Figure 2.21. *Capacitive effect caused by*
the electric field between two conductors

Figure 2.21 helps understand the appearance of the capacitive coupling between two lines, thanks to the lines of the electric field between the conductors. The capacitors C and Cf represent the capacitance toward the mass and the edge effect, respectively. An increase in the coupling capacitance leads to a decrease in the parasitic capacitance toward the mass. This mutual capacitance is inversely proportional to the distance between the two tracks.

– Magnetic field creating an inductive effect between the tracks;

As shown in Figure 2.22, when two conductors are close, the magnetic field created by the track in which a current is circulating influences the second track by inducing a current. The interaction coefficient between the tracks is the mutual inductance that results in a phenomenon comparable to an electrical transformer.

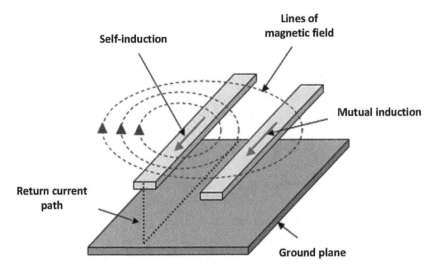

Figure 2.22. *Capacitive coupling and inductive effects*
caused by the magnetic field between two conductors

Throughout the following theoretical calculations, we have only considered inductive coupling, and this is the predominant coupling for tracks of a FR4-type printed circuit board. Moreover, as ESD discharges are current sources, magnetic fields are predominant compared to electric fields.

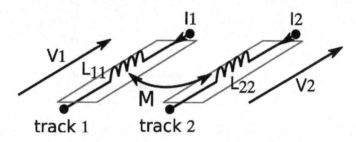

Figure 2.23. *Representation of the electric model, including the inductive coupling of the two lines*

The inductive coupling of two microstrip lines is represented in Figure 2.23. Coefficient M is the mutual inductance between the tracks. V1 and V2 are the voltages induced in tracks 1 and 2, respectively.

The system made up of two lines is a quadripole whose matrix of the inductances is written as [BAK 90]:

$$\begin{bmatrix} V1 \\ V2 \end{bmatrix} = \begin{bmatrix} L_{11} & L_{12} \\ L_{22} & L_{21} \end{bmatrix} * \begin{bmatrix} \dfrac{dI_1}{dt} \\ \dfrac{dI_2}{dt} \end{bmatrix} \qquad [2.17]$$

The inductances L_{11}, L_{22} are the inductances of the lines and L_{12}, L_{21} are the mutual inductances between the tracks. In the case of a symmetrical system, the tracks are identical: $L_{11} = L_{22}$ as well as $L_{12} = L_{22} = M$ (Figure 2.23).

The diagram of the principle behind the measurement system carried out is illustrated in Figure 2.24(a). The equivalent electrical diagram of the system is provided in Figure 2.24(b). Source V_S of impedance Z_s, modeling the TLP pulse bench, is connected to track 1. A load impedance Z_C, at the end of the line, represents the impedance of a load that could be an ESD protection. Two impedances, Z_a and Z_b, terminate track 2. L_1 and L_2 are the line inductances and pM is the source of the magnetic force corresponding to the coupling in the Laplace domain.

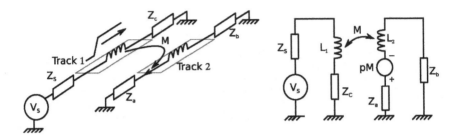

Figure 2.24. *Schematic representation of the measurement system (a) and of the equivalent electrical system (b)*

The laws of Kirchhoff and Ohm's law provide the values of the voltages V_a, V_b at the terminals of Z_a and Z_b, and the expression of current I_1 in the Laplace domain (p being the complex Laplace variable):

$$V_a(p) = \frac{Z_a}{Z_a + Z_b + Z_{L2}}(pMI_1) \text{ and } V_b(p) = \frac{-Z_b}{Z_a + Z_b + Z_{L2}}(pMI_1) \tag{2.18}$$

with $I_1 = \dfrac{V_s}{(Z_s + Z_c + Z_{L1})}$ [2.19]

Following these equations, the voltage induced at the terminals of Z_a and Z_b in the victim track corresponds to the derivative of the current that travels through the disturbing tracks, multiplied by the factor M. By neglecting the value of the self-inductance of lines L1 and L2 and by combining equations [2.18] and [2.19] in the time domain, the expressions of V_a and V_b become:

$$V_a = \frac{Z_a}{Z_a + Z_b} M \frac{\partial I_1}{\partial t} = \frac{Z_a}{(Z_a + Z_b)(Z_s + Z_c)} M \frac{\partial V_s}{\partial t} \tag{2.20}$$

$$V_b = -\frac{Z_b}{Z_a + Z_b} M \frac{\partial I_1}{\partial t} = -\frac{Z_b}{(Z_a + Z_b)(Z_s + Z_c)} M \frac{\partial V_s}{\partial t} \tag{2.21}$$

This is therefore the same as deriving the signal Vs and applying an amplification factor to it. Consequently, the magnetic coupling taking place between the lines causes a disturbance in line 2, whose amplitude is dependent only on the rising and falling edges of the stress signal. This results in a positive and a negative square signal, resulting from the rise time (Tr) and the fall time (Tf) of the pulse, as shown in Figure 2.25. Following the expressions of V_a and V_b, it can be noted that their waveforms are inverted at each extremity of the victim line. The appearance of a rectangular signal is linked to the initial approximation, which neglected L1 and L2, giving equations for the induced voltages corresponding to first order derivatives. By adding the self-inductances of the lines into the equations, which are of the same order as the mutual inductance, the equation of the induced voltages increases by an order and as a result, the form of the generated signal is more complex.

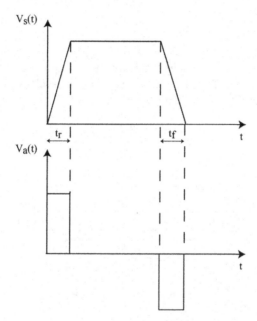

Figure 2.25. *Effect of inductive coupling on a pulse*

To summarize, the induced voltage is maximal during current transitions. In our case, we are looking to use this system to carry out measurements. In order to maximize the induced voltage at the terminals of Z_A, the victim line must end with a short-circuit, meaning that $Z_B = 0$ and $V_B = 0$. The impedance Z_A can be considered equal to the oscilloscope impedance of 50 Ω.

From the equations [2.18] and [2.19], the system transfer function is written as:

$$F(p) = \frac{V_a(p)}{V_s(p)} = \frac{Z_a}{(Z_a + Z_b + Z_{L2})(Z_s + Z_c + Z_{L1})}(pM) \qquad [2.22]$$

By assuming the system to be symmetrical, the impedances of the lines are then identical: $Z_{L1} = Z_{L2} = Z_L$. Moreover, by considering the line terminations to be purely resistive, the transfer functions can then take on the following form:

$$F(j\omega) = \frac{j\dfrac{\omega}{\omega_0}}{\left(1 + j\dfrac{\omega}{\omega_1}\right)\left(1 + j\dfrac{\omega}{\omega_2}\right)} \qquad\qquad [2.23]$$

With $\omega_0 = \dfrac{R_a(R_S + R_C)}{M.R_a}$ $\qquad \omega_1 = \dfrac{R_a}{L} \qquad \omega_2 = \dfrac{(R_S + R_C)}{L}$ $\qquad\qquad [2.24]$

The last few expressions help show that the coupling is dependent on the frequencies **f0, f1** and **f2**. The frequencies are themselves dependent on the termination impedances and the line impedances.

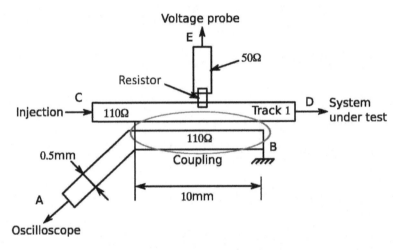

Figure 2.26. *Example of the implementation of inductive coupling on a printed circuit board*

Based on this, it is possible to easily develop a pattern on the printed circuit board that allows the current to be measured. A diagram representing the test pattern is shown in Figure 2.26. The circled area corresponds to the coupled lines. The tracks used have a width of 0.5 mm (characteristic impedance of 110 Ω). They are separated by a distance of 0.25 mm. The coupling takes place over a length of 1 cm. The voltage probe presented in Figure 2.18 (section 2.2.1) for the TDR method is integrated into the pattern (port E). This is used during the calibration operations that are presented in the following section.

2.2.3.2. Current reconstitution

Whichever the magnetic probe used, or on the base of the pattern presented previously, the current must be reconstituted mathematically. The calculation can be performed in one of the two ways: using an integral method or using a frequency method.

Integral method

For this method, a first approximation must be considered, which is that the coupling is perfectly linear in terms of frequency (+20 dB/dec). Equation [2.18] allows the following expression of the current I_1 to be established:

$$I_1(t) = \frac{Z_a + Z_{L2}}{Z_a} * \frac{1}{M} * \int V_a(t) * dt \approx \frac{1}{M} * \int V_a(t) * dt \qquad [2.25]$$

With Z_b equal to zero and by holding that $Z_a >> Z_{L2}$. The coefficient k = 1/M is determined through experimentation using a calibration pattern, or using analytical formulae, depending on the type of structure for which the coupling is taking place [BAK 90, DEM 06, BOY 07]. Note that in formula [2.25] the coupling coefficient k, which is the mutual inductance M, must be calculated or extracted by precise calibration in order to obtain an accurate value of the current.

There are two possibilities: the coupling takes place from track to track on the printed circuit board as represented in Figure 2.26 (this implies that a coupling pattern has been implemented on the board). Loops from close magnetic fields can also be used, as shown in section 2.2.3.4.

In the case where the coupling structure from Figure 2.26 is used, the calibration pattern must have the same geometry. Only a footprint is added at the end of line 1, allowing a resistor to be placed, a short-circuit to be made, or for the circuit to be left open. Calibration takes place in the following manner:

– A resistor of 110 Ω is connected to the end of line 1 (port D, Figure 2.26) allowing the impedance to be adapted and to avoid any reflections onto the line.

– A TLP of 100 ns, with transition times of 1 ns is injected into port C (Figure 2.26).

– A measurement of the transmitted wave is carried out with the integrated voltage probe (port E, Figure 2.26). The current traveling through is calculated using the expression:

$$i(t) = \frac{V_{transmitted}(t)}{110} * a \qquad [2.26]$$

where α is the attenuation coefficient of the integrated voltage probe.

– A measurement of the voltage induced in the coupled track is carried out in port A. The coupling coefficient k is determined from equations [2.25] and [2.26] using the expression:

$$k = \frac{\sum_{t=t1}^{t2}(\frac{V_{transmitted}(t)}{110})}{\sum_{t=t1}^{t2}(\int V_{probe}(t) * dt)} \qquad [2.27]$$

In the same way as a quasi-static TLP characterization (described previously in this chapter), the coefficient is obtained by calculating the mean value between the times t1 and t2, which corresponds to the plateau of the TLP pulse sent out. Once the coupling coefficient k has been determined, the current traveling through track 1 (Figure 2.26) is calculated from the voltage in port A (Va) using equation [2.25].

Frequency method

The approximation made using the integral method is only valid for low frequencies. Beyond certain frequencies, the frequency response of the probe is no longer linear (Figure 2.29), and another method is required for calculating the value of the current in the frequency domain. The coupling coefficient is then determined by carrying out a convolution using the frequency measurement. The fast Fourier transform (FFT) algorithm is used to calculate the discrete Fourier transform (DFT). This allows a function represented in the time domain to be transformed into a function represented in the frequency domain and *vice versa*. The Fourier transform is applied for periodic signals. In our case, the disturbances sent are impulsive. In order to apply FFT, we must verify that the signals measured go back to zero, meaning that they can be considered periodic.

As for the integral method, the coupling coefficient in the frequency domain is determined through experimentation using a calibration pattern. In the example provided in Figure 2.26, a resistor of 110 Ω is connected at the end of line 1 (port D), allowing the impedance to be adapted, and avoiding any possible reflections along the line. A TLP pulse of 100 ns with a transition time of 1 ns is injected into port C. The injected current is written as Ie(t), and its discrete transform as Ie(k). The discrete transform of the induced voltage Va(t) measured at the terminals of the probe is Va(k). The coupling coefficient is given through the relation:

$$Coef(k) = \frac{Ie(k)}{Va(k)}$$
[2.28]

Once the coupling coefficient Coef(k) is determined, the induced current flowing through the coupled track (or in the probe) is calculated from the product of the DFT of the voltage measured in A (Va) using the coupling coefficient:

$$I(k) = Coef(k) * Va(k)$$
[2.29]

The inverse transform allows a move back into the time domain.

2.2.3.3. Experiments and validation

In order to validate the two "integral" and "frequency" calculation methods, the following section offers a presentation of the experiments carried out using the calibration method presented in Figure 2.26. A TLP pulse of 10 V of 100 ns, with a transition time of 1 ns, is sent onto port C. The result of the current calculation is given in Figure 2.27. The transmitted current obtained using the voltage probe (port E) is superimposed over it in the same figure.

This current has a value of 90 mA when the terminal impedance is adapted (110 Ω). In the case of a short-circuit, the value of the transmitted current is maximal. This is due to the addition of the more reflected incident current. In the case of an open circuit, the whole current is reflected, and the current is equal to zero. We can see in the results from the current calculations that this behavior is well reproduced.

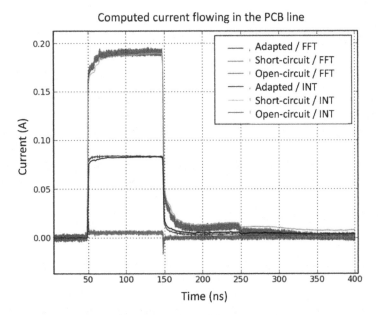

Figure 2.27. *Transmitted current measured in port E and the current traveling through the track calculated using the integral method and the frequency method (adapted track termination with a short-circuit and an open circuit). For a color version of this figure, see www.iste.co.uk/bafleur/esd.zip*

2.2.3.4. *Current measurement using near-field magnetic probes*

The diaphonic measurement method allows the ESD currents traveling along the tracks of a board to be measured using a specific pattern. This method can therefore not be used for a final electronic product. The use of near-field magnetic probes allows the dynamic surface currents to be measured for any electronic board.

An electromagnetic field is made up of an electric component $\vec{E}$ and of a magnetic component $\vec{H}$. In order to make the measurement easier, it must be possible to consider these two components separately from each other. Looking at the radiation of antennae, different zones can be distinguished, including "near" field, and "far field" zones. The separation between these zones is defined as a function of the wavelength λ. If the observation distance L from the source is L<λ/2π, the conditions are considered to be of "near-field" conditions. One component is then predominant over the other. Thus, for a monopole antenna, the electric field component is predominant. The field is said to be of "high impedance". Inversely, for a current loop type antenna, the magnetic field is predominant, and the field is said to be of "low impedance". The components of the electric and magnetic fields are no longer in phase, and their relation becomes more complex. However, fields $\vec{E}$ and $\vec{H}$ can be measured independently in this zone depending on the type of antenna used. For this reason, we used near-field conditions to measure the magnetic fields emitted by the tracks of the printed circuit board, using a current loop.

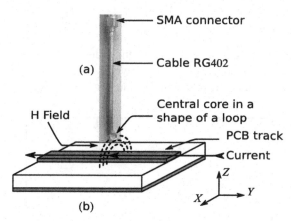

Figure 2.28. *Photograph of the near-field magnetic probe (a) and measurement principle (b)*

The principle behind the field measurement using the probe is illustrated in Figure 2.28. Magnetic field $\vec{H}$ is made up of three components Hx, Hy and Hz. The orientation of the loop allows measurement of the field following one component, in particular. The current loop is sensitive to field lines that are perpendicular to its surface. In the example presented here, the current loop is located in the ZY plane. The measured component of the magnetic field is Hx. When the field goes through the loop, a value of the current proportional to the value of the field is induced in the inductance of the loop, creating a potential difference. According to [DEM 06], the voltage measured using small probes as a function of the size of the loop is written as:

$$V(j\omega) = -j\omega.\mu_0.\frac{\pi D^2}{8\pi r}.I(j\omega)$$ [2.30]

with I being the current induced in the loop, D the diameter of the loop, r the distance from the loop to the track and $\mu_0 = 4\pi \times 10^{-7}$ the magnetic constant.

A measurement of the frequency response of the probe is shown in Figure 2.29. This measurement is carried out using a spectrum analyzer and a sinusoidal generator. The near-field probe, 3 mm in diameter, is placed 0.3 mm away from a 50 Ω track ending with an impedance of 50 Ω. By applying equation [2.30], we can determine the coupling coefficient, which is the value of the mutual inductance, M, from the parameters used during the measurement:

$$M = \mu_0 \cdot \frac{\pi D^2}{8\pi r}$$
[2.31]

From these measured fields, and thanks to the value of the mutual inductance extracted through calibration, the current traveling along the tracks can be recalculated, using either the integral method or the frequency method.

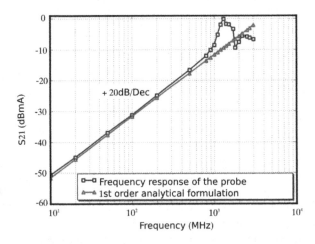

Figure 2.29. *Frequency response of a near-field measurement probe (loop diameter of 2.5 mm) and extracted first order analytical formula*

The following two sub-sections describe the building of the probe and the experiments carried out using it, respectively.

2.2.3.5. Dynamic current circulation mapping

Part three of standard IEC 61967 defines a method for measuring the electromagnetic radiation emitted by a circuit by carrying a near-field scan above the package or the chip itself [IEC 14]. A passive field probe is moved above the system under test, which allows a map to be made of the near-field radiation. The measurement and predictions of the radiated fields have been extensively studied and published in the literature [AIM 07, AOU 08]. The method put forward in this section reuses the principle of near-field scanning defined in standard IEC 61967-3, to which is added the method of current reconstruction presented previously. The goal is to create a map of the current traveling through the system undergoing an ESD event from the fields measured. Similar measurement techniques were first proposed and published in the literature by [FUK 04], and then in parallel by [MON 11a], [CAI 10] and [HUA 10].

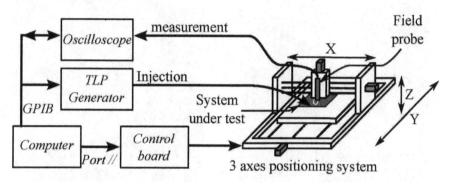

Figure 2.30. *Simplified schematic representation of the automatic characterization set-up for near-field scanning used in current reconstruction*

A simplified schematic representation of the automatic measurement set-up is provided in Figure 2.30. At the center of the system is a 3-axes positioning machine. The injection (TLP generator) and measurement (oscilloscope) devices are connected to the electronic board under test and to the field probe, respectively. In order to automate the measurement, these devices and the control board of the machine are controlled using a computer. For each step of the probe, the TLP pulse injected into the board under test is repeated, and the voltage induced in the probe is measured and recorded in a file.

The step size that the probe moves along by defines the resolution of the final images. As we have explained, magnetic probes can only measure along a single field orientation $\vec{H}$. In order to reconstitute the field, the two components Hx and Hy must be obtained. This is done by pivoting the probe by 90°. The resultant of the field is obtained by summing the vectors of the fields Hx and Hy. The intensity of the field is determined for each acquisition point by calculating the modulus following the relation:

$$V_{resultant} = \sqrt{V_{Hx} + V_{Hy}}$$

[2.32]

This expression allows the creation of an image of the evolution of the total field emitted by the current flow according to the components Hx and Hy, as illustrated in Figure 2.31. These images correspond to the field emitted during the propagation of the current along a bent track at different moments of the acquisition (resolution of 1 mm).

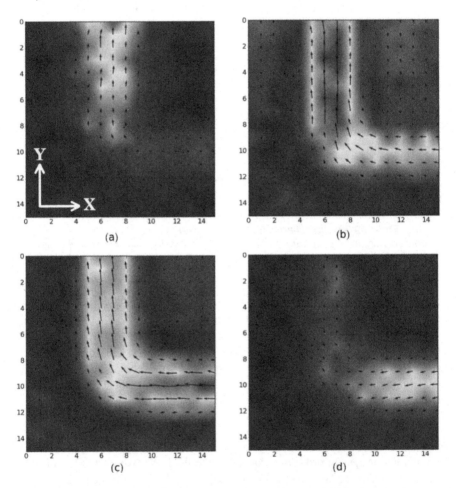

Figure 2.31. *Distribution map of the emitted magnetic field for successive and close measurement samples (a), (b), (c) and (d). For a color version of this figure, see www.iste.co.uk/bafleur/esd.zip*

The arrows correspond to the resultant of the field. The current in the tracks travels from the top to the right in the track with a 90° bend. At the start of the transition (Figure 2.31(a)), the field starts to become established in the vertical track.

During the transition (Figure 2.31(b) and (c)), the field is maximal and is emitted by the two segments of the line. Finally, at the end of the transition (Figure 2.31(d)), the intensity of the current is constant and no longer varies. The field is then only emitted by the horizontal track.

From the resultant fields obtained by the probe following components Hx and Hy, the distribution of the current traveling through the track can be traced out for each couple (X,Y) in the acquisition zone. The distribution map of the current obtained using the integral calculation method for the bent track is provided in Figure 2.32.

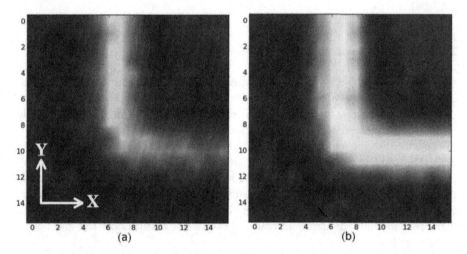

(a)   (b)

Figure 2.32. *Distribution map of the current in the bent track, calculated using the integral method, at two different instants: appearance of the current (a) and constant current (b). For a color version of this figure, see www.iste.co.uk/bafleur/esd.zip*

This mapping technique for currents flowing through boards is an investigation tool that allows analysis of the propagation of currents within electronic systems according to the protective topologies used. This measurement technique is put into practice in the case study presented in section 5.3 of Chapter 5.

2.3. Injection methods

In order to test for failures, ESDs must be injected into the systems. Two problems arise, which are the choice of stress, and the manner in which they are injected, which must follow a configuration that is as realistic as possible. With regards to the choice of stress, given that the IEC 61000-4-2 is the only existing standard to date, it is *de facto* the official reference. However, as seen in Chapter 1, the gun presents a number of problems itself, linked to radiation and reproducibility. Moreover, the generated waveform is relatively complex and makes investigations aimed at better understanding the phenomena taking place in the system impossible. As a result, we prefer to use the TLP, which generates square signals, which are entirely controllable and configurable.

Injection can take place through contact, called conduction mode, or by radiation (radiation mode). In the literature, injection methods by radiation are used to create electromagnetic disturbances in order to study susceptibility [LAC 07b, LAC 08].

2.3.1. Injection in conduction mode – DPI method

Part 4 of the IEC 62132 [IEC 07b] standard defines an injection method used to evaluate the immunity of a circuit with regards to conducted electromagnetic disturbances caused by the injection of a voltage in a pin of the circuit. The principle of injection relies on the use of a capacitor to transmit a high frequency (HF) disturbance directly into the pin of the circuit in operation, as illustrated in Figure 2.33. The capacitor allows the HF generator to be isolated from the circuit. By progressively increasing the power of the harmonic stress until a failure is induced, the immunity of the component can be estimated. A series resistance can be added in order to limit the current. Decoupling networks are placed onto the supply signals to make sure that the HF signal introduced disturbs the circuit under test and not its power supply.

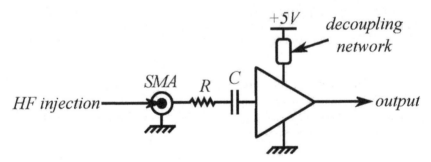

Figure 2.33. *Principle of the DPI (Direct Power Injection) injection method [IEC 07b]*

This method can be modified [LAC 07a, ALA 08] in order to inject a TLP or IEC-type ESD disturbance into an operating electronic system. The disturbance is sent into the supply track, as illustrated in Figure 2.34, or into an input/output of the component. A decoupling network must be inserted onto the supply pin, so that the disturbance only stresses the circuit under test, and not the power supply, which can have a low impedance compared to the impedance present in the integrated circuit.

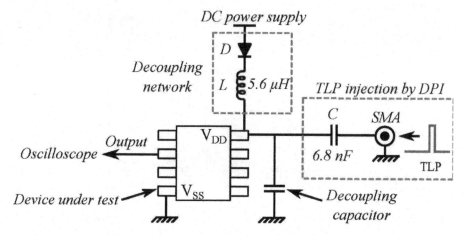

Figure 2.34. *Diagram representing the injection method of TLP by DPI with a power supply of VDD, creating a disturbance in an operating circuit [MON 11a]*

This injection technique is put into practice and described in detail in Chapter 5 in order to analyze the susceptibility of a flip-flop D installed as a frequency divider. As we shall see, the disturbance injected onto the supply pin can induce clock cycle losses.

2.3.2. Near-field injection

As we have already stated, the near-field mapping method, as referenced in standard IEC61967-3 [IEC 05b], is used to capture the electromagnetic fields radiated by a circuit under test in operation. Figure 2.35 presents the principle behind the characterization bench for the near-field emissions of a circuit in the case of electromagnetic compatibility measurements (EMCs). The voltage induced in the probe, which is proportional to the amplitude of the components of each field captured, is measured by the spectrum analyzer and provides information on the amplitude of the field for a classical EMC measurement. A high-precision positioning table allows the probe to be moved at constant distance from the surface under test. Automated positioning systems, including positioning and data acquisition programs, provide precision and repeatability in the order of one micrometer.

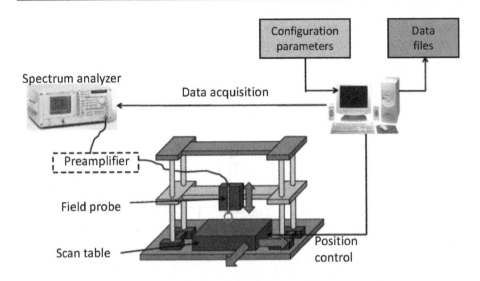

Figure 2.35. *Principle behind the near field scanning measurement set-up [BOY 07]. For a color version of this figure, see www.iste.co.uk/bafleur/esd.zip*

The principle behind the test bench is the same as that followed during emission, the difference being that an amplified harmonic signal is sent toward the circuit under test through an injection device, which in this case is the near-field probe. This immunity measurement technique, which allows injecting stresses into integrated circuits or electronic boards, presents several advantages. On the one hand, it is a radiation method that does not present any particular constraints in terms of the setup of the circuit under test in the electronic board, as no modifications are required for the disturbances to be injected. Moreover, Boyer showed that the influence of the magnetic probe, which is used in our studies, on the behavior of the DUT is negligible [BOY 07]. Furthermore, the disturbance is produced locally by injecting electromagnetic fields into a narrow area, such as the pins of the circuit package. Work carried out by S.G. Zaky [ZAK 92] and J.J. Laurin [LAU 91] using disturbances induced by a miniature loop at the level of the terminals of a voltage-controlled oscillator revealed modifications in terms of the operating frequencies of the oscillator.

Although the near-field probe has mainly been used in studies of harmonics, Wang and Pommerenke have carried out localized electrostatic discharge injections to create susceptibility maps of electronic boards [WAN 04b, LAC 06]. This method allows for a better understanding of the influence of the radiation of a transient source on the electric activity of a DUT. Thus, the near-field method, compared to the methods of disturbance by direct conduction, provides new information on the susceptibility of circuits regarding radiated transient disturbances.

A normative document regarding this method was edited by ESDA, ANSI/ESD SP14.5-2015 – Near Field Immunity Scanning – Component/Module/PCB Level [ANS 15b]. The stress injected by a near-field probe is a square pulse with a rise time of 1 ns and a far longer fall time, meaning that only one edge participates in the generation of a fault. An example of the application of this technique in the analysis of the susceptibility of a microcontroller is presented in case study 4 (section 5.4).

2.4. Failure analysis techniques

Many analysis techniques help localize problems such as short-circuits or open circuits following ESD issues. The main characteristic of ESD failure signatures is their size, which usually have a diameter of 10 µm or less, except for certain events, such as those that follow a CBM (Charged Board Model) type discharge [OLN 03]. Not all techniques of fault localization can be reviewed here, and as such we shall focus only on the main ones. The simplest technique, and the most obvious one for large faults is examination using optical microscopy. We shall not discuss this well-known technique in order to better look at those that are used to analyze faults that it does not reach.

2.4.1. Static light emission microscopy (EMMI)

Light EMission MIcroscopy is a technique for direct fault localization that has been used since the end of the 1980s in front-side analysis [KHU 86]. Later, it was improved in order to be used in back-side analysis [KAS 98]. It involves observing the light emitted by a component during operation in a failure configuration. Certain microscopic defects, such as current leakage at junctions, metallic protrusions, MOS transistors that are saturated when the current is flowing, avalanche breakdowns at junctions, the phenomenon of latch-up, leakage currents located in oxides, filaments of polysilicon and some damages of the substrate, can cause physical phenomena that result in the emission of light.

Two main families of photons can be distinguished:

– photons generated during collisions between carriers accelerated by an electric field: leakage current in a reverse-biased diode, reverse-biased diode (avalanche diode), bipolar transistor in a non-saturated state, leakage current in the oxides and MOS transistor during saturation;

– photons generated by the radiative recombination of the electron-hole pairs: direct diode, bipolar transistor in the saturated state, thyristor in the on-state and latch-up.

The damage caused by ESD stress in a reverse junction and in dielectrics (gate oxides) results in leakage currents that lead to the emission of light, thus making EMMI a technique that is well-adapted to this type of defect [WIL 88].

In order to locate the defect(s), the component is biased in a configuration of failure, and is placed under a microscope with a camera specific for detecting photoemissions. The bias voltage can be increased to improve the efficiency of this technique, although this does increase the risk of changing the nature or the size of the defect. When coupled to a pulse generator like the TLP, this technique allows observation of the behavior of a protection component [RUS 98], thus providing additional information regarding its function at various current levels. The TLP must be repeated at a certain frequency to obtain a big enough contrast, which is often around 10 Hz, over several minutes (1–10 min), so that the photoemission camera can gather enough photons. Note that there is not temporal information, since the photons are integrated throughout the entire length of the capture. Figure 2.36 presents EMMI images of a CMOS oscillator circuit (two different samples) having undergone 10 successive CDM-type ESD stresses of 2 kV [GUI 04a, GUI 06]. The defects observed correspond to, on the one hand, a short-circuit occurring between the drain and the substrate of the PMOS transistor (Figure 2.36(a)), and on the other hand, a breakdown of the gate oxide of the PMOS transistor (Figure 2.36(b)).

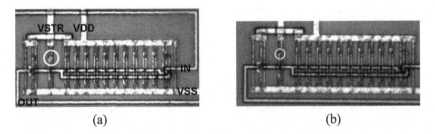

(a)   (b)

Figure 2.36. *EMMI localization of an ESD defect in CMOS oscillator: the image at the top corresponds to a short-circuit between the drain and the substrate of the PMOS, and the bottom image to a breakdown of the gate oxide of the PMOS transistor*

2.4.2. Dynamic light emission microscopy (PICA)

Light emission microscopy, or EMMI, when coupled with a TLP tester, helps obtain information regarding the behavior of an ESD protective network, for various current levels, for example. However, it does not provide any temporal information on the emission observed due to the necessary integration of the signal over several minutes, needed to get a usable image.

Spatial and temporal information can be acquired alongside light emission, thanks to a specific detector. A new technique called PICA ("Picosecond Imaging Circuit Analysis") uses dynamic light emission to detect the emission induced during switching of a CMOS gate [KAS 99]. When a transistor goes from the off-state to the on-state, or vice versa, it transitions momentarily through a state of saturation. In this state, the transistor emits light when a current flows through it. The photons are then generated by radiative recombination of the hot carrier induced by the strong electric field associated with this type of conduction. Visualization of these light emissions over time allows the spatial and temporal propagation of an electrical signal to be followed. The PICA technique is therefore able to identify functionality issues by front-side or by back-side for current circuits, with a time resolution in the order of a few picoseconds [MCM 00].

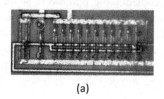

(a)

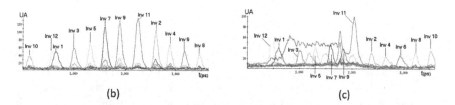

(b)   (c)

Figure 2.37. *PICA dynamic light emission used with a front-side CMOS oscillator: (a) defect localization (circle) by dynamic EMMI, temporal emission spectra for the reference piece (b) and for the piece that has undergone cumulative CDM stress (c). For a color version of this figure, see www.iste.co.uk/bafleur/esd.zip*

This technique is usually used in the industry as a probe, checking whether two signals in the center of the circuit have not become desynchronized, an electrical signature that can provide the source of a failure or of the presence of a defect. This technique can provide more information on the defect detected than the classical photoemission technique. Figure 2.37 presents a dynamic EMMI image for a CMOS oscillator component having undergone cumulative CDM stresses and compares its PICA temporal emission spectrum with that of a reference component [GUI 06]. On the spectrum of the reference piece, it can be noted that only switching from the high level to the low level is visible, which explains why at the start only the switching of uneven stages can be seen, followed by the switching of even stages during the following oscillation period. This is due to the difference between ionization coefficients [TRA 84] of the PMOS and NMOS transistors, which is far greater for the latter, thus resulting in light emission that is far greater than that of PMOS. With regard to the piece placed under stress, the oscillator stage *Inv11*, which undergoes the ESD stress directly, emits during the entire semi-period where the uneven inverters switches. This is linked to the presence of defects at the level of the PMOS transistor of this stage (Figure 2.37(a)), which disturbs the behavior of the NMOS, and therefore of its emissions. During this period, the NMOS gate of this inverter is at the low level, the oxide breakdown defect in the PMOS causes a leakage current through its gate, which increases the potential of the input of that stage. As the gate of the NMOS is no longer at 0 V, it is in saturated conduction, which is the operating mode that causes considerable emissions at the level of the NMOS. A more in-depth study of these results is presented in the thesis by M. Remmach [REM 09]. The technique of picosecond imaging (PICA) highlights the presence of a disturbance at the stressed gate, but does not provide precise information regarding the exact location of this/these defect(s). In complex circuits, it is a tool that can help isolate the failing block before moving onto techniques for localizing the failing element.

2.4.3. *Laser stimulation techniques*

Of all the techniques commonly used for precisely localizing defects, laser stimulation techniques, which use a laser beam to interact with the integrated circuit being analyzed, have experienced the most remarkable success and development. These techniques are usually applied to the back-side of the circuit to avoid the many metallic levels of the front-side of modern circuits.

One category of these techniques that has been successfully used for localizing ESD defects is the methods that use a laser as a source of disturbance, with the energy deposited being low enough so as to not damage the circuit. The laser beam interacts locally with the various materials that make up the integrated circuit. Two effects, induced by photoexcitation, are mainly used, with the wavelength of the laser used for determining the nature of the excitation obtained:

– heating (photothermal effect) or thermal laser stimulation (TLS);

– the generation of electron-hole pairs by photon absorption (photoelectric effect) or photoelectric laser stimulation (PLS).

A wavelength in the order of 1.3 μm is usually used for thermal stimulation, while photoelectric generation is carried out with a wavelength of around 800 nm.

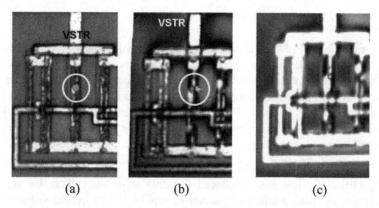

(a)   (b)   (c)

Figure 2.38. *Comparison of the various localization techniques for detection of a resistive defect induced by CDM type ESD stress: (a) EMMI, (b) OBIRCH and (c) SEI. The stressed sample is the same for images (a) and (b). For a color version of this figure, see www.iste.co.uk/bafleur/esd.zip*

2.4.3.1. *Thermal laser stimulation (TLS)*

TLS techniques are well-adapted to the localization of resistive defects. The different TLS techniques are the following: OBIRCH or Optical Beam Induced Resistance Change [NIK 98], TIVA for Thermally Induced Voltage Alteration [COL 99] and SEI for Seebeck Effect Imaging [COL 99]. These different techniques are all based on the same principle. A laser with a wavelength in the order of 1.3 μm locally heats the circuit studied. Measurement of the current (OBIRCH) or of the voltage (TIVA) at its terminals, in function of the position of the laser, provides a map of the variations measured. The presence of a defect generates a signature difference and makes localization possible. For OBIRCH and TIVA techniques, the circuit is biased, while in the case of the SEI technique, the circuit is not. The voltage variations measured in this case are caused by the Seebeck, or thermocouple, effect: heating of two different materials creates a potential gradient. This last technique presents the advantage of not aggravating or modifying the nature of the defect during analysis, since no bias is applied to its terminals. These techniques allow efficient localization of defects [ESS 06] inducing current intensities in the order of the microampere, which are difficult to be detected using the photoemission technique (Figure 2.38). Defects inducing currents of less than 1 μA remain hard to detect, however.

2.4.4. Photoelectric laser stimulation (PLS)

The principle of PLS lies in the photogeneration of surplus electron/hole pairs through the absorption of photons by silicon. The interaction between the laser, which has a wavelength in the order of 800 nm, and the chip is more complex than in the case of thermal stimulation. As for thermal stimulation, the technique lies in the detection of the variation of electrical characteristics at the terminals of the component, under the excitatory effect of the laser. A map of the variations measured is carried out by scanning the surface of the component to be analyzed. As for photoelectric laser stimulation, three different techniques can be identified:

– the OBIC (Optical Beam Induced Current) technique, where the electrical value analyzed is the variation of the power supply current, as the voltage of the power supply is kept constant [WIL 86];

– its dual technique, LIVA (Light Induced Voltage Alteration), which involves measuring variations of the power supply voltage, with the current of the supply being maintained constant;

– the NB-OBIC (Non-Biased OBIC) technique, for which the component is not biased and only the current created by photogeneration is measured [KOY 95, BEA 03]. The amplitude of the electrical signal obtained is dependent on the intensity of the electrical field in the area of generation of the carriers caused by the excitation of the laser. The electrical field causes the free carriers generated to become separated and causes a photocurrent to appear. The bigger the electrical field, the greater the current. The amplitude of the photocurrent also depends on the rate of recombination of the carriers in the area of generation.

It has been shown that the same defect leads to an increase of the photocurrent in the OBIC mode, and a decrease of the photocurrent in the NB-OBIC mode [ESS 04]. The sensitivity of this technique allows for the characterization of small defects that can be the cause of leakage currents in the order of one nano-ampere. The defects were previously undetectable using photoemission and thermal laser stimulation techniques.

Figure 2.39 illustrates the detection of latent defects caused by cumulative HBM stress in a bipolar NPN transistor, inducing a leakage current of less than a µA [GUI 05]. The defects were undetectable using EMMI and OBIRCH techniques. In these images, the zones in red are the metallic zones of the collector and the emitter, and between the two there is the metallurgic base-collector junction. In the reference images (a) and (b), the map of the photocurrent is homogenous for that zone. However, for the stressed component, images (c) and (d), two zones can be observed, lighter for OBIC and darker for NB-OBIC, which correspond to silicon microfilaments that short-circuit the junction of the base-collector.

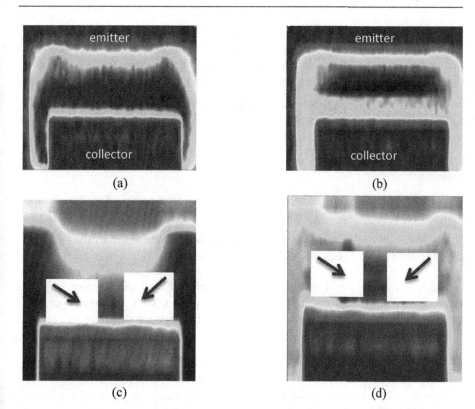

Figure 2.39. *Localization of latent defects using OBIC and NB-OBIC techniques in a bipolar NPN transistor undergoing cumulative HBM type ESD stresses: (a) NB-OBIC image and (b) OBIC image of the reference component, (c) NB-OBIC image and (b) OBIC image of the component having undergone the ESD stress. The red arrows mark the defective areas. For a color version of this figure, see www.iste.co.uk/ bafleur/esd.zip*

2.4.5. Detection of latent defects by low frequency noise measurements

Defects caused by ESD stress are usually very small in size (<1–10 μm). The base technique for detecting the presence of a defect is to measure the leakage current, either at the input or output protection that has been stressed, or by measuring the quiescent current of the circuit, or I_{DDQ}. However, ESD defects generate a weak leakage current that can prove to be undetectable in a complex circuit, especially if the defect was induced within the circuit as a result of a failure of the ESD protection strategy. For applications that require long-term reliability, such as aeronautical, space or automotive applications, it is important to be able to detect possible latent defects that could reduce this reliability.

A technique that we have explored is the measurement of low frequency noise [GUI 04b]. The measurement of low frequency noise involves recording the frequency spectrum of the current going through the structure for a given bias. Figure 2.40 presents the measurement set-up that we used. It is made up of:

– a resistive biasing system, the resistance must be as high as possible (10 MΩ) for the thermal noise linked to this resistance to be low enough so as to not mask the source of the noise being measured;

– a decoupling capacitor, which eliminates the continuous component of the signal;

– a transimpedance amplifier, which amplifies the fluctuations of the current and transforms them into voltage fluctuations;

– a spectrum analyzer (HP 8941A), which provides a transform of the temporal signal so as to bring it back into the frequency domain.

Considering the low levels of the signals involved, the measurement of low frequency noise (LF) is carried out in a Faraday cage, with a power supply by batteries for the transimpedance amplifier and for the biasing system.

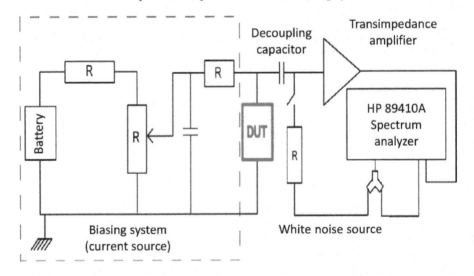

Figure 2.40. *Low frequency noise measurement set-up*

The measurement of low frequency noise (LFN) describes all undesirable disturbances within the electrical system that are superimposed over the useful signal and tend to mask its contents. Thanks to the measurement of the frequency of this disturbance [BAR 01], all electrical noise occurring in the form of random and spontaneous voltage and/or current fluctuations caused by different physical processes provides a large amount of information in the low frequency domain (less than 1 MHz).

By identifying the various sources of noise, it is possible to obtain information on the physical defects of the components or of the materials. This information can be used to detect defects with high levels of sensitivity and/or optimize manufacturing or technology, especially for qualifying the dielectric or RF products. The various sources of noise mainly belong to four types: diffusion noise, shot noise, 1/f noise and G-R (generation-recombination) noise.

Diffusion noise is linked to the interaction of electrons with atoms of the crystal lattice. It exists even in the absence of an electric field and is equivalent to thermal noise. It is the minimum amount of noise generated by a perfect sample and its amplitude is directly proportional to the temperature.

Shot noise only exists when an electric current is present. It results from the flow of carriers through a potential barrier, such as the one induced by a junction.

G-R noise is associated with the process of generation and recombination of carriers in semiconductor components. It is linked to the presence of defects in the semiconductor, which is reflected in the existence of an energy level, the occupation of which fluctuates over time, affecting the number of free carriers either by recombination or by generation of these carriers.

1/f noise is characterized by the fact that the power of noise is inversely proportional to the frequency, as the name suggests. It is caused either by the fluctuations of the number of carriers due to G-R over several traps simultaneously at the surface and at an interface, or by fluctuations of the mobility of the carriers associated with large amounts of electron-phonon collisions.

In order to evaluate the different components of noise, a simplified model of the spectral density of the noise is usually used and described by the following equation:

$$S_i(f) = A + \frac{B}{f} + \sum_i \frac{C_i / f_{ci}}{1 + \left(f / f_{ci} \right)^2} \quad \text{with } f_{ci} = 2\pi\omega_i \qquad [2.33]$$

where the first term A corresponds to white noise (2qI+4kT/R), the second term B corresponds to 1/f noise and the last term represents the sum of the G-R noise, which corresponds to the presence of one or several generation/recombination centers or to traps located in the oxide and/or at the Si/SiO$_2$ interface. f_{ci} corresponds to the different capture and emission frequencies of the charges at the various centers of generation/recombination.

Figure 2.41(a) presents the results obtained for a GGNMOS-type ESD protection, submitted to successive TLP stresses inducing a very slight degradation of the I-V characteristic of the structure, with an increase in the leakage current at a voltage of 6 V, going from 40 to 500 pA. Such a slight degradation can pass unnoticed during industrial qualification tests. This structure was also characterized through LF noise measurement, Figure 2.41(b), after each stress level. For stress level no.10, the generated ESD defect induces an increase of nearly two decades for the measurement of the noise, so a far greater level of sensitivity than for the measurement of the leakage current. In Chapter 5, we will present a case study in a complex circuit, and we will show that in such a particular case, only the LF noise measurement technique is able to detect the defect.

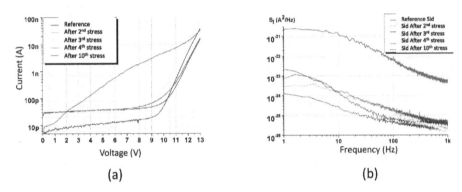

(a)   (b)

Figure 2.41. *Electrical characterization of a GGNMOS protection submitted to successive TLP stresses: (a) measurement of the leakage current and (b) measurement of low frequency noise. For a color version of this figure, see www.iste.co.uk/bafleur/esd.zip*

2.5. Conclusion

In order to design components and electronic systems that are able to resist ESDs, it is vital to have adapted investigation tools. The base characterization technique is the TLP measurement bench, which since its creation has changed to adapt itself to, on the one hand, new standards such as the CDM with VF-TLP, and on the other hand, to the requirements of advanced CMOS technologies and electronic systems on boards.

TLP generators are essential tools for characterizing ESD protection structures in quasi-static states. Used in reflectometry, they are useful investigation tools for the validation of line models under pulse stresses. This diversity of use makes it a key tool for characterizing phenomena as much at the component level as at the system level.

Coupled with failure localization techniques based on laser stimulation, these tools allow validation of the different discharge paths of an integrated circuit or of an electronic board with the goal of minimizing the design iterations tied to ESD protection strategies. These techniques are also vital for building compact or behavioral models of protection strategies.

3

Protection Strategies Against ESD

In the Introduction, we mentioned the various precautions that are taken in order to protect components and electronic systems from ESD. When the system is in its application environment, it no longer benefits from all these precautions, and protective structures must be put in place to protect it from potential ESD stresses.

In this chapter, we would first of all like to recall the impact of the evolution of integrated circuit technology on ESD robustness. We will then review the various protection strategies, from integrated circuits to electronic board systems. After introducing the major design constraints of these protections, we shall describe the elementary devices (both bipolar and MOS) that allow implementation of these protection strategies and provision of efficient protection. At the system level, we shall see that the addition of discrete protection to the electronic board, the passive components and routing must be taken into consideration during the optimization of the protection strategy.

3.1. ESD design window

The efficiency of a protective component against electrostatic discharges is defined by its ability to protect an integrated circuit no matter the combination of pins under stress or the discharge polarization. The protective component must constitute a preferred path during the discharge and it must be able to absorb the entirety of the energy of the ESD event without degrading it, while limiting the voltage generated over all of the nodes of the circuit.

The design of an integrated circuit's protection against ESD is influenced by large constraints in terms of cost, as the components, which are absolutely vital in terms of the robustness of the circuit, do not participate to improve functionality, or degrade it through associated parasitic effects. Thus, these components must usually be designed without any additional technological steps and must occupy as small a silicon area as possible. These protections are normally placed at the inputs/outputs of the integrated circuit and often below the bonding pads to limit their surface footprint. The surface available for ESD protection circuits is therefore limited, with an imposed shape factor.

As these protections are not used when the integrated circuit is operating normally, they must not change the functionality of the circuit, meaning that they must be entirely transparent and behave like an open switch. As a result, the leakage current and parasitic capacitance of an efficient protection component must be limited and must in no case be activated when the circuit is operating normally.

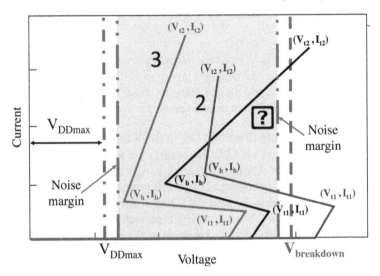

Figure 3.1. *ESD design window. For a color version of this figure, see www.iste.co.uk/bafleur/esd.zip*

In order to fulfill these requirements, an ideal ESD protection component must have its current and voltage characteristics located within a design window, an example of which is provided in Figure 3.1. The lower threshold of this window is equal to the power supply voltage, plus a safety margin of 10%, to make sure that the normal operation of the circuit is not disturbed. The upper limit is defined by the breakdown voltage of the weakest element of the technology node (oxide breakdown, junction breakdown, etc.), also with a margin of 10%.

In Figure 3.1, we report several I-V characteristics of protection structures presenting negative resistance, which is typical of components built from bipolar transistors or thyristors. In order to conform with the design window, the protection trigger voltage, called V_{t1}, must be greater than the nominal power supply voltage of the circuit V_{DD} +10% (safety margin). The current level required for the protection to be triggered is I_{t1}. Despite these precautions, it is possible for a protective structure to be triggered unexpectedly during normal operation of the integrated circuit following injection of parasitic current into the substrate. In CMOS integrated circuits this phenomenon is called "latch-up", corresponding to the triggering of a parasitic thyristor. The protection circuit must then be able to return to the state of open switch while the circuit is being powered. To do this, the holding voltage of the protection trigger, written V_h, must be greater than the power supply voltage, and its holding current I_h must be big enough. Finally, the current I_{t2} and the voltage V_{t2}, which correspond to the failure of the protective system, define the intrinsic robustness of the protection.

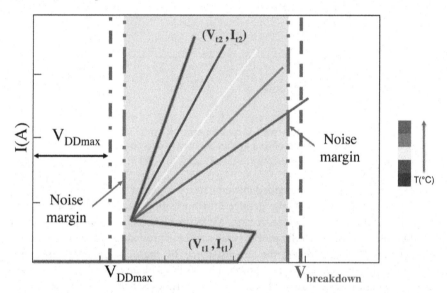

Figure 3.2. *Impact of temperature on the ESD design window. The color code for the temperature on the right of the graph shows blue as low temperatures and red as the highest temperatures. For a color version of this figure, see www.iste.co.uk/bafleur/esd.zip*

In Figure 3.1, the protections whose characteristics are curves 1 and 2 cannot protect an electronic circuit, since the first one has a high on-resistance, which induces a premature breakdown, and the second one is triggered for a voltage V_{t1} greater than the breakdown voltage. The ideal protective structure is therefore one whose electric I-V characteristics correspond to curve 3.

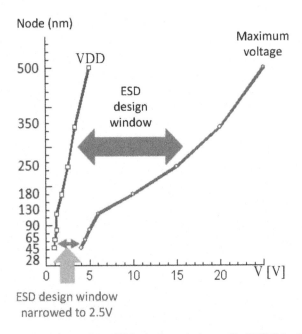

Figure 3.3. *Evolution of the ESD design window with CMOS technology nodes. For a color version of this figure, see www.iste.co.uk/bafleur/esd.zip*

There is a tendency in embedded systems to try to operate electronic equipment at higher temperatures to achieve greater miniaturization and to reduce weight. In integrated CMOS circuits, the use of large MOS transistors is very common to carry out ESD protection between the power supply and the ground rail, usually called the power clamp (PC). However, the on-resistance of these components increases greatly with the temperature. For example, a MOS PC with an on-resistance of 4.4 Ω and an ESD failure current of 1.74 A at 25 °C undergoes a change in robustness and on-resistance at 200 °C, to 1.34 A and 7.1 Ω, respectively [ARB 14]. As shown in Figure 3.2, such a degradation of performance can cause the protection to no longer comply with the design window, meaning that the required level of ESD robustness is not supplied when the circuit is operating at high temperatures [ARB 12]. It is therefore important to design ESD protections that take this additional parameter into account in order to provide effective protection under all operating conditions.

Beyond these various constraints, this design window is also greatly influenced by the evolution of micro and nano-scale electronics. The reduction of transistor size according to Moore's Law [MOO 65] is concomitantly accompanied by a decrease in the supply voltage and a reduction of the breakdown voltages of these advanced technologies. This results in a considerable reduction of the ESD design window, as shown in Figure 3.3. This evolution has resulted in a review of ESD robustness specifications, under the impulse of the consortium *Industry Council* [IND 13], which advised for a reduction in the HBM robustness of these technologies, from 2 to 1 kV. This reduction was justified by the analysis of a large amount of customer returns from the manufacturers, which showed that the ESD robustness of a system was not at all correlated with the intrinsic robustness of its components [SME 08]. This unexpected result shows the importance and the necessity of a global approach in terms of system ESD protection.

Beyond these constraints of having to conform to an increasingly narrow design window, the protection structure must also comply with other requirements:

– tolerance to variations in the process, considering that for more complex technology, this variability is more pronounced;

– robustness to several successive ESD stresses: for example, in automotive applications, a component can undergo several thousand successive stresses over its lifetime [RIV 04], while the HBM qualification standard only requires the application of three successive stresses;

– protection for various ESD models (HBM, MM, CDM, HMM);

– no interaction with circuit operation: the component must have a minimal level of parasitic capacitance, particularly for RF circuits, as well as a leakage current that is as weak as possible. For low noise applications, the parasitic noise must be minimal;

– pass the different reliability tests, such as the latch-up test, immunity to which must be taken into consideration during design;

– immunity to electrical transients, and electromagnetic interference (EMI) in particular. An important trend in the design of the circuits and systems involves adopting a shared approach for protection strategies against ESD and EMI [ABO 11];

– to be able to protect the circuit both when powered and when not powered.

3.2. Elementary protective components

The first step in protecting integrated circuits against ESDs is to control the environment in which they are manufactured (level of humidity, ionization of the air, antistatic floors and work surfaces, etc.), with the establishment of an area that is protected against ESD, called the EPA. In addition to this EPA zone, several specific measures can be taken (antistatic wrist straps, antistatic bag or box for transporting the components, for example) with regard to the handling of electronic components in order to reduce the occurrence of ESD events. However, these precautions can only be taken during manufacturing; when the components leave the EPA areas they must be able to protect themselves. This implies the implementation of specific ESD protective structures onto the silicon, which significantly improves the robustness of the semiconductor components, greatly reducing failures caused by ESD.

3.2.1. Protection strategies

The protection strategy against ESD involves integrating protective structures onto the chip that provide an effective discharge path for the ESD current, while limiting the voltage experienced by the internal components of the integrated circuit. This mainly means protecting the dielectrics and junctions of these components, the breakdown voltages of which have decreased significantly following Moore's Law (Figure 3.3), resulting in an increasingly narrow ESD design window. This efficiency is needed both in terms of the current intensity (or ESD robustness) and of the surface occupied by the silicon (and therefore cost).

The protection strategy relies on elementary components that we shall describe in section 3.2.2 to propose a current path for the discharge, no matter the number of pins on which the ESD stress is being applied. Without polarization and without a clear path to the ground, the ESD current can indiscriminately enter and exit along any pin. Depending on the application and on the technology involved, the protection strategies can quickly become complex (the number of pins being an important factor in the increasing complexity).

Three protection strategies can be differentiated: the localized protection approach, the so-called "centralized" approach and the distributed approach. Each of these has advantages and disadvantages, and the choice depends on the type of circuit to be protected (analog or digital), the technology (advanced CMOS or mixed technology), the complexity and size of the chip as well as the cost.

The localized protection strategy (Figure 3.4(a)) involves placing ESD protection structures at each pin that requires protection. The structure must be bidirectional in order to provide protection from the two polarities of the discharge. Its main advantage is that it simplifies the architecture and the placement of the ESD protection structures by implementing a shared ground bus all around the chip. This approach also helps limit any coupling between the different pins.

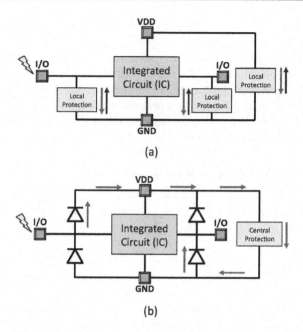

Figure 3.4. *Different approaches for ESD protection:*
localized (a) and centralized (b). For a color version
of this figure, see www.iste.co.uk/bafleur/esd.zip

While remaining localized, this approach must allow for the existence of a discharge path that is effective for all types of discharge. For example, in Figure 3.4(a), a discharge between the two I/O pins would involve the local protections linked with these pins, one in direct conduction mode and the other in the reverse.

The disadvantage of this approach lies in the need to develop different types of protection structure for each type of input/output and each power supply pin.

The inputs of a CMOS technology integrated circuit in particular require an efficient limitation of the voltage to protect the gates of the MOS transistors, which are relatively sensitive. A typical protection of CMOS inputs is the Π protection [DUV 89], which is illustrated in Figure 3.5. It consists of a primary and a secondary protection, both isolated by a resistor. The primary protection is designed to absorb most of the ESD current, while the secondary protection is meant to trigger the primary protection and ensure that the voltage is limited to a low enough level. The resistor limits the current toward the circuit being protected. In this way, when the I/O contact undergoes an ESD stress the secondary protection is triggered first, allowing the ESD current to pass through toward the ground. This current also travels through the resistor R and causes an increase in the voltage on the side of the primary protection, until it is triggered itself. For the protection to be efficient, the impedance of this primary protection must be far smaller than that of the secondary protection. A primary protection can be created using diodes or an SCR. The resistor is usually made of polysilicon, which has the advantage of inducing small parasitic elements. For the secondary protection, a GGNMOS component (described later) is often used. If the technology contains a protection that is robust, with a low trigger voltage and a low on-resistance, the primary protection can be sufficient, but in most cases the resistance is preserved in order to limit the current.

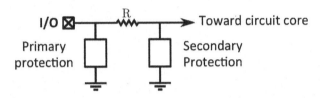

Figure 3.5. *Diagram of localized Π protection of a CMOS input*

This design work must be accompanied by a concurrent effort of surface optimization of each structure, in an attempt to reduce the size of the chip and therefore the cost. Moreover, localized protective structures are usually developed from parasitic components of the technology (bipolar transistors, thyristors), and require specific models to be developed in order to simulate the protection strategy at the level of the chip.

Finally, due to the necessary bi-directionality of the protection system, the surface – and therefore the parasitic capacitance associated with the structure – can be incompatible with some applications operating at very high frequencies.

The centralized protection approach is particularly useful in this context. It involves using diodes at the inputs/outputs in order to redirect the ESD current toward one or several centralized protections installed at the level of the power supply pins, as illustrated in Figure 3.4(b). The diodes only work in direct mode, making them very robust, and only requiring a small silicon area. This protection approach is therefore well adapted for high-performing circuits. The design effort is mainly focused on centralized protection structures, power clamps, at the level of the power supply pins, as well as the routing of the ground bus and power supply bus.

This strategy presents the advantage of considerably reducing the ESD design time, including modeling, since the components used are part of the library of active components of the technology involved. The diodes only require an extended characterization under a strong current and in pulse mode.

However, this technique is limited in terms of the number of power supply voltages in a same chip (more than three). Indeed, the complexity of the architecture of the metallic buses needed increases with the number. Moreover, this technique cannot be used for isolated power supply pins.

In order to deal with these constraints, a distributed protection approach was developed, as presented in the following publications [STO 03, STO 04, STO 05]. The principle behind it involves distributing all the protections as close as possible to the input/output contact points in order to minimize the access resistance or even to place all of the protections in the same cell (under the same contact point). When a discharge occurs between two inputs/outputs, a trigger circuit detects it and activates all of the protections of the contact points of the circuit. This helps maintain a low voltage in the power supply buses and reduces the size of the protections. In this way, it is possible to protect circuits containing a high number of pins. An example of a distributed protection strategy is presented in Figure 3.6. Each contact point is made of three diodes and of a NMOS transistor. The gray line marks the path of the current when a discharge takes place between point 1 and point 2. During this discharge, diode D1 biases the "Boost" rail, allowing the ESD to be detected by the trigger circuit. The trigger circuit in turn biases the "trigger" rail. The current is distributed throughout the NMOSs and is evacuated by point 2. Thus, the voltage drop in the bus remains low.

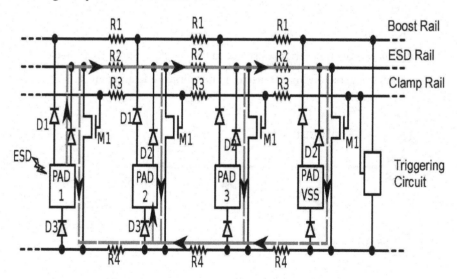

Figure 3.6. *Distributed ESD protection approach*

3.2.2. Elementary protection structures

Although the active components of an integrated circuit are able to self-protect under certain conditions, thanks to an appropriate design, this is not usually the case, and specific ESD protection systems must be put into place.

The role of the ESD protection structure is to provide an efficient discharge path for the ESD current without damaging the internal components of the integrated circuit. The protective structure must therefore be robust against the discharge against which it is meant to protect, which means resisting intense electric fields and dissipating strong current densities (from 10^4 to 10^6 A.cm^{-2}). As a result, a vertical protective component is naturally more robust than a lateral component due to the greater volume of silicon available for the dissipation of energy. However, the performance levels of the protective component are very dependent on its geometry and on the technology it is used in.

Under normal operating conditions of the integrated circuit, the protective structure must disturb the functionality of the circuit as little as possible and must ideally behave as an open switch whose main characteristics are (Figure 3.7):

– having a high value impedance;

– having a weak leakage current;

– inducing a low parasitic capacitance;

– bringing a minimal in series resistance to the contact point being protected;

– not being triggered during a transient in the normal operating mode.

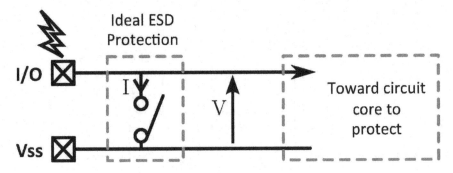

Figure 3.7. *Principle of an ideal ESD protection*

However, in the presence of an electrostatic discharge, the protective structure should ideally behave as a closed switch, with the following characteristics:

– very low impedance;

– very high switching speed (≤ 1 ns for an HMB stress ≤ 100 ps for a CDM stress);

– limiting the voltage applied to the terminals of the component to protect;

– robustness against various types of ESD stress (HBM, CDM and IEC) and against cumulative stress over the course of the entire lifespan of the circuit;

– passing the different qualification tests (beyond the ESD tests) such as the latch-up test, for example;

– being unaffected by the fluctuations of the technology parameters, knowing that for cost reasons there is usually no specific component that has been developed to create these protective structures. However, some technological steps can be developed to increase performance. One example involves the localized blocking of silicides in the drain area of an MOS transistor in order to improve the efficiency of the ballast technique [OH 02].

Finally, the silicon area occupied by these protections must be minimized in order to reduce its impact on the final cost of the integrated circuit being protected. The robustness levels achieved are expressed in V/μm and the efficiency in V/μm^2. In order to reduce their silicon footprint, some technologies allow the protections to be placed under bonding pads, resulting in significant gains.

3.2.2.1. Diodes

At the start of the 1980s, the diode was the most commonly found protection structure. It is still often used, especially in the centralized protection strategy. Its simple structure does not contain any thin dielectric films, which tend to be sensitive to ESD, thus providing it with good intrinsic robustness.

In forward bias mode, diodes provide very efficient protection as they dissipate very little energy. This efficiency is also seen in a very low level of parasitic capacitance. Their robustness in terms of current density can be adjusted by modifying the surface. However, the trigger voltage (~0.5 V for a PN diode) cannot be adjusted, although several diodes can be cascaded in order to increase the overall trigger voltage.

Reverse-biased diodes enter conduction by an avalanche mechanism or through the Zener effect. The avalanche takes place when the electric field locally reaches a critical value beyond which the carriers have enough energy to multiply by ionizing the atoms of the crystal lattice by impact. The breakdown voltage depends on the doping of the N and P regions that form the junction. The diode is fragile in this conduction mode as its resistance in the passing state is usually quite high and the current density is normally not very homogenous.

Strongly doped N^+-P^+ junctions break down through the Zener effect. In such a scenario, the extension of the space-charge zone is so small that the carriers go directly from the valence band to the conduction band by tunnel effect. The Zener diodes then present a lower value of resistance than the diodes in avalanche mode. Their breakdown voltage is lower, typically around 6–7 V, which is very useful for limiting the voltage over the gates of the MOS transistors, for example. However, the value of their leakage current during normal operation is often higher, as is the value of their parasitic capacitance.

3.2.2.2. Bipolar transistors

After the diode, the bipolar transistor is the most used component for ESD protection. This protection is either an active component in the technology (as in smart power technologies), or a parasitic component as in the case of CMOS technologies. It is usually the bipolar NPN transistor that is preferred, but in some cases it can be better to use a PNP transistor, as we shall see.

When the bipolar NPN transistor is used in its self-biased mode (Figure 3.8(a)), its base is linked to the emitter, either directly through a short-circuit or by using a resistor. When a negative discharge is applied to the collector, the emitter being taken as the reference electrode, the transistor behaves likes a forward-biased diode through its collector/base junction. This case is very favorable to the dissipation of the discharge current as long as the value of the series resistance of the diode is not too large.

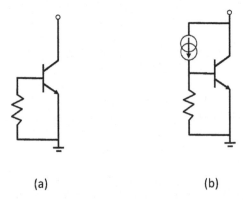

(a)   (b)

Figure 3.8. *Bipolar NPN transistor configured for ESD protection: self-biased mode (a) and bias by an external current supply (b)*

In the case of a positive discharge onto the collector with respect to the emitter, there are two methods available to trigger the bipolar transistor. When configured in self-biased mode (Figure 3.8(a)), it is the avalanche current of the collector/base junction that is used. The other solution involves an external current supply between the electrodes of the collector and of the base (Figure 3.8(b)). In this case, the current supply is provided by another transistor (bipolar, MOS) or by a reverse-biased diode, usually a Zener diode. This configuration makes it possible to adjust the trigger voltage of the transistor. The I_B current thus generated flows throughout the base resistor, R_B. When this current reaches a threshold, such that:

$$I_B . R_B = V_{BE} \text{ where } V_{BE} = 0.5V \tag{3.1}$$

it induces the triggering of the bipolar transistor. The discharge current for which this relation is verified is the trigger current of the ESD protection, which is often called I_{t1}.

Figure 3.9 presents the measured TLP characteristic of a self-biased bipolar transistor, used in smart power technology. This will help us in illustrating the operating principle of the protection structure. It can already be seen that as long as the breakdown voltage of the collector–base junction has not been reached, there is no current flowing through the structure. At the trigger point (I_{t1}, V_{t1}), the emitter injects electrons that contribute to the phenomenon of the multiplication of carriers in the space-charge zone of the collector–base junction, triggering the bipolar effect. The resulting increase of the collector current I_C, and therefore of the emitter current I_E leads to a decrease in the avalanche multiplication coefficient M.

The simplified implicit equation governing this snapback phenomenon is the following:

$$\alpha \cdot M = 1 + \frac{I_B}{I_E} \tag{3.2}$$

where α is the gain of the common-base bipolar transistor (self-biased configuration), M is the avalanche multiplication coefficient, I_B is the base current and I_E is the emitting current. M is a function of the collector–base voltage according to the empirical Miller's formula [MIL 57]:

$$M = \frac{1}{1 - (\frac{V_{CB}}{BV_{CB}})^n} \tag{3.3}$$

with $2 < n < 6$ and BV_{CB} being the breakdown voltage of the collector–base junction.

According to equation [3.3], a decrease in the multiplication coefficient M also induces a reduction in the voltage required at the terminals of the structure to maintain the level of avalanche current, and this results in a snapback of the I-V characteristic of the transistor, point (I_H, V_H), which allows the structure to dissipate the discharge energy while minimizing heating.

The I-V electrical characteristic of a self-biased bipolar transistor is therefore the result of the coupling between two current multiplication phenomena: the bipolar effect and the avalanche.

The increase of the current level then leads to a strong modulation of the conductivity in the base, which helps achieve a low value of on-resistance, and provides the component with good ESD robustness, especially when it is set up as a vertical structure.

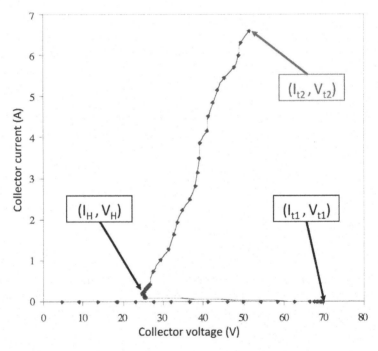

Figure 3.9. *Measured TLP characteristic of a self-biased bipolar NPN transistor used in smart power technology. The couple (I_{t1}, V_{t1}) defines the trigger point of the structure, (I_H, V_H) defines its snapback point and (I_{t2}, V_{t2}) the failure point. For a color version of this figure, see www.iste.co.uk/bafleur/esd.zip*

When this self-biased bipolar transistor is used to protect a high-voltage pin, this snapback of the characteristic, which is beneficial for the ESD robustness of the protection structure, can be problematic as the value of the snapback voltage V_H can be lower than the value of the supply voltage, and therefore not conform to the ESD design window.

In order to control the value of the holding voltage V_H, the more commonly used techniques involves cascading as many bipolar transistors as necessary to conform to the ESD design window. The disadvantage of this solution is the need to significantly increase the area of each component in order to maintain the same global on-resistance of the protection. This increase in the silicon area can be prohibitive in terms of production costs, and for this reason there has been a lot of research conducted looking for alternatives. One solution involves modifying the structure of the bipolar component, for example by increasing the resistance of the collector using various techniques that modulate the doping of the buried layer [GOS 02].

Another alternative involves changing the bipolar component, or in other words using a bipolar PNP transistor instead of an NPN. As shown in equation [3.2], in order to increase the weight of the multiplication coefficient M, and therefore to increase the associated voltage that defines V_H, it is best to reduce the gain of the bipolar transistor, which is intrinsically what happens with bipolar PNP transistors. This type of transistor also has a lower multiplication coefficient as it is linked to the hole carriers. As a result, when in a self-biased configuration, the characteristic of the PNP transistor undergoes very little or no snapback. We therefore only need to adjust the trigger voltage V_{t1}. These transistors also have a lateral structure that, when combined with the properties cited previously, results in a value of the on-resistance that is intrinsically greater than the resistance of an NPN transistor. Several combined solutions have been suggested to resolve this problem and have been proven to be effective [GEN 07a]:

– create an inter-digitized and multi-finger architecture, with a minimum dimension for the emitter and the collector, and no ballast resistance;

– remove the base contact, since the PNP component presents little or no snapback. This allows the total silicon area of the component to be reduced significantly (up to 25% gain);

– as far as the technology allows it, use an abrupt doping profile for the collector, and low doping levels for the base, in order to favor high-injection effects.

In smart power technologies like the BCD (Bipolar/CMOS/DMOS), it can be beneficial to take advantage of the parasitic vertical PNP component in order to improve the performance of this component [GEN 07b]. Figure 3.10(a) illustrates such a component. The parasitic vertical PNP is simplified here to a diode as its gain is nearly equal to zero due to the presence of the buried layer. Optimization of this component requires adjustment of the trigger voltage of the two components in parallel so that they are triggered simultaneously. In the component shown here, this is obtained by adjusting the lateral distance D between the N type zone and the anode diffusion P$^+$. The second parameter to be controlled is the density of the current in each component, which defines the global on-resistance of the protection. This parameter would be optimized using the length L, which defines the surface of the vertical diode. The TLP characteristic of the optimized PNP structure, in comparison to a standalone PNP structure (Figure 3.10(b)), shows a significant improvement in performance, with an on-resistance of 1 Ω and an HBM robustness of 8 kV HBM for a silicon area of 10,000 μm^2.

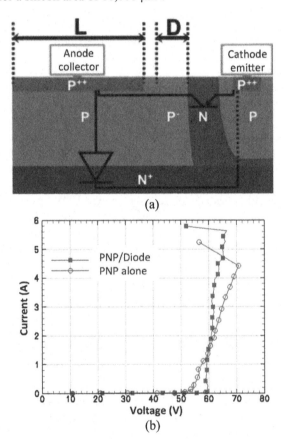

Figure 3.10. *Schematic cross-section of a PNP transistor coupled with the parasite vertical diode (a) and TLP characteristic compared to the standalone PNP component (b). For a color version of this figure, see www.iste.co.uk/bafleur/esd.zip*

3.2.2.3. MOS transistor

In advanced CMOS technologies, the MOS transistor (P type or N type) is the only component available. It can be used either in MOS operation or in association with its parasitic bipolar component. Three cases can arise:

– self-protection of an output level: use of the parasitic bipolar to provide partial ESD protection;

– localized protection: use of the parasitic bipolar to provide partial ESD protection;

– centralized protection: MOS operation alone.

3.2.3. Self-protection of an output level

To illustrate our point, let us consider the case of a simple input/output stage such as the one presented in Figure 3.11(a). It is made up of an input, which is a simple inverter, and an output, which is also a simple inverter, made of a PMOS transistor and a NMOS transistor. In order for this input/output level to self-protect from ESD, the designer relies on the parasitic components associated with this stage. They are grayed out in the electrical diagram of this figure. At the level of the NMOS, there is an N-type lateral parasitic bipolar, or LNPN, comprising the source (emitter), the drain (collector), and the P-substrate of the CMOS technology (base). Furthermore for the PMOS, there is a P-type lateral parasitic bipolar, or LPNP, comprising the source (emitter), the drain (collector) and N-well of the CMOS (base) of the technology. There is also a P-type vertical parasitic bipolar, or VPNP, comprising the drain (emitter), N-well of the CMOS technology (base) and the P-substrate of the CMOS technology (collector). The diagram also shows the drain/substrate diodes D_{P+} and D_{N+} associated with each MOS transistor, as well as the parasitic diode made up of all of the N wells with the P-substrate, D_{NWell}, as well as its associated parasitic capacitance C_{NWell}. These parasitic components (except the diodes) are represented in the schematic technological cross-section of Figure 3.11(b).

The parasitic components are particularly efficient in the case of ESD stress involving forward-biasing of these components. This is the case for the following three ESD stresses:

– positive stress on I/O relative to VDD (Zap A in Figure 3.12): it is diode D_{P+} that provides a path for the ESD current by forward biasing itself;

– negative stress on I/O relative to VSS (Zap B in Figure 3.12): it is diode D_{N+} that forward biases itself;

– positive stress on VSS relative to VDD (Zap C in Figure 3.12): it is diode D_{NWell} that forward biases itself.

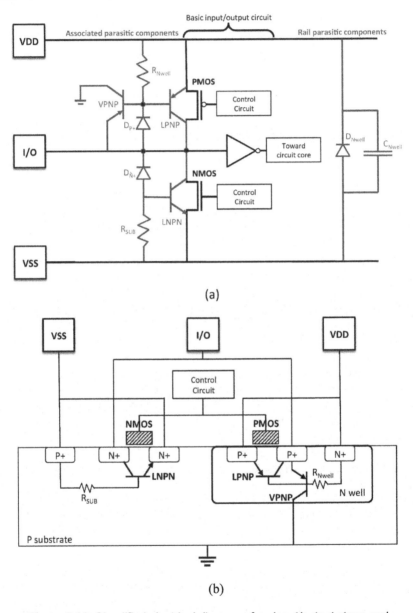

(a)

(b)

Figure 3.11. *Simplified electrical diagram of an input/output stage and of the associated parasitic components (a) and schematic cross-section of CMOS technology with parasitic bipolar components (b)*

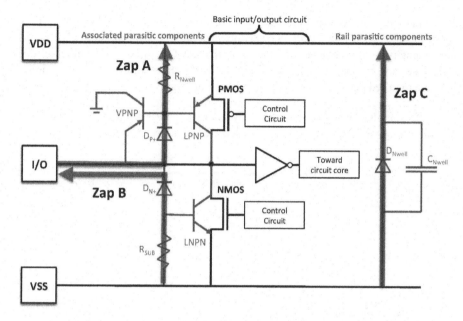

Figure 3.12. *ESD current paths in three modes of ESD stress involving forward-biasing of parasitic diodes: Zap A positive on I/O relative to VDD, Zap B negative on I/O relative to VSS and Zap C positive on VSS relative to VDD*

For the other biases of ESD stress, the paths are a bit more complex and the robustness is less guaranteed.

– Starting with positive stress on VDD relative to VSS:

If the value of the parasitic capacitance C_{NWell} is big enough, i.e. in the order of 10–80 nF (only true for big chips), the ESD current can be absorbed by this capacitance. In this case, it is necessary to have very robust metallizations. Usually, however, this capacitance is not big enough, and a specific protection regarding this biasing must be implemented.

– Positive stress on I/O relative to VSS (Zap D, Figure 3.13):

The main pathway for this polarity is the conduction of the parasitic bipolar LNPN transistor associated with NMOS, thanks to the avalanching of the diode D_{N+}.

This conduction mode of the bipolar transistor is susceptible to the formation of hot spots and there is a risk of the current focalization, which is pushed away by the implementation of a ballast resistance in the NMOS drain. The value of this resistance, and as such of the associated ESD robustness, must be a compromise with the required performance levels of this output stage.

There can also be a secondary pathway (more resistive) going through the diode D_{P+} associated with the PMOS and through the capacitance C_{NWell}.

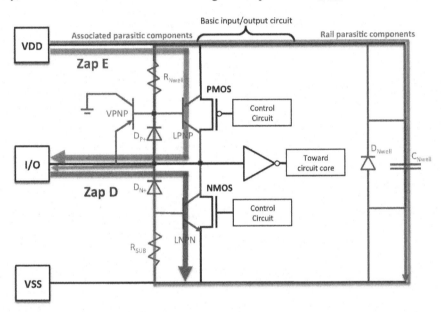

Figure 3.13. *Current paths for two ESD stress configurations between I/O and power supplies: positive Zap D on I/O relative to VSS (red paths) and negative Zap E on I/O relative to VDD (blue paths)*

– Negative stress on I/O relative to VDD (Zap E, Figure 3.13):

This is the symmetrical opposite of the previous case. The main path for this case is the conduction of the parasitic bipolar LPNP transistor associated with the PMOS, thanks to the avalanching of diode D_{P+}. The implementation of a ballast resistance in the PMOS drain is also required to ensure a certain level of ESD robustness.

There can also be a secondary pathway (more resistive) through the diode D_{N+} associated with the NMOS and through the capacitance C_{NWell}.

– The last case is illustrated in Figure 3.14 and involves an ESD stress between two input/output pads:

The main path for this type of stress (positive here on the contact point I/O_1) is the conduction of the parasitic bipolar LNPN transistor associated with the $NMOS_1$ of the left output stage, thanks to the avalanching of diode D_{N+} and the forward-biasing of the diode D_{N+} associated with the $NMOS_2$ of the right output stage.

There can also be a secondary path (shown as a thinner line) through the diode D_{P+} associated with the $PMOS_1$ of the left output stage and through the capacitance C_{NWell} and the forward-biasing of the diode D_{N+} associated with the $NMOS_2$ of the right output stage.

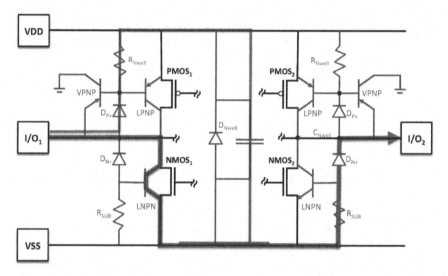

Figure 3.14. *ESD current paths for an ESD stress between two input/output pads. The ESD stress has a positive bias on the I/O1 pad relative to the I/O2 pad*

3.2.4. Localized protection

When self-protection is not possible in CMOS technology, as in the case of the input to a logic gate or an amplifier, the first approach involves implementing a form of localized protection at the pad that needs to be protected. As in the self-protection of an output stage, the principle involves using the parasitic bipolar component of an NMOS transistor optimized for ESD protection.

The simplest configuration for this protection is called GGNMOS (Gate Grounded NMOS), which means using an MOS transistor, usually a NMOS, set up in diode configuration (Figure 3.15(a)). In the case of a negative discharge, the drain-substrate diode forward-biases and provides an efficient discharge path. When the discharge is positive, the voltage must reach the avalanche breakdown value of the drain–substrate junction in order to allow a great enough current to flow through the substrate (base of the parasitic bipolar NPN transistor) and then the triggering of the parasitic bipolar NPN transistor.

In order to trigger this parasitic bipolar, the GCNMOS (Gate Coupled NMOS) uses the conduction of the MOS transistor by allowing a coupling of the ESD stress to take place on the gate of the MOS (Figure 3.15(b)). It is the value of the resistor R that allows us to adjust this coupling, as the capacitance C can be the gate-drain capacitance or an additional capacitance of higher value. This second option is usually used to create a clamp between power supply rails VDD and VSS. The graph in Figure 3.15(c) shows the significant impact of this setup on the trigger voltage of the structure, which goes from more than 14 V in the GGNMOS setup (R=0) to 9 V for a resistance of 140 kΩ.

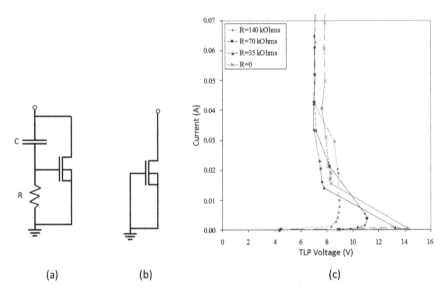

(a)   (b)   (c)

Figure 3.15. *NMOS transistor-based ESD protection: GGNMOS (a) and GCNMOS (b). Measured TLP characteristic showing the impact of the value of the resistor R on the trigger voltage in the GCNMOS setup GCNMOS (c) [BER 01]. For a color version of this figure, see www.iste.co.uk/bafleur/esd.zip*

The evolution of CMOS technologies, resulting in the feature shrinking and density integration that exist today, has also caused a great reduction in the supply voltage and breakdown voltages of gate oxides. However, these self-biased components need the junctions to be biased up to the level of their breakdown voltage by avalanche, whose value is then close or equal to the value of the breakdown voltage of the gate oxides. Furthermore, ever since the 0.25 μm CMOS generation, the bipolar LNPN component – which is a major element in the localized ESD protection strategy – has become very fragile and particularly sensitive to technological fluctuations, meaning that it no longer ensures the proper ESD protection required. This protection approach therefore cannot be used for advanced CMOS technology. In the current sub-micron CMOS technology, a centralized protection approach is preferred.

3.2.5. Centralized protection

We presented this protection approach in Figure 3.4(b), which involves re-directing the ESD currents onto a central protection using diodes. In the case of CMOS technologies, this central protection is established using a MOS transistor from the library. This allows for an easy simulation of the protection strategy, since the electrical model is available but must be large in order to absorb the ESD currents. The easiest solution to implement is a GCMOS, as presented in Figure 3.15(b), which is ideal for simple circuits. A classic type of implementation for more complex integrated circuits is presented in Figure 3.16. The protective MOS transistor is triggered dynamically through a RC circuit that activates a chain of inverters [MER 93], hence its name of active ESD clamp. Only the MOS operation is activated in this configuration, which is why the transistor must be big, so as to absorb ESD currents of several amperes. The time constant of the RC circuit must be chosen so as to be longer than the duration of the ESD event, but far shorter than the power ramp, which is in the order of several milliseconds. An ESD event with transients in the range of several nanoseconds, for example an RC constant of 1,000 ns, would help the dynamic circuit avoid discriminating an ESD event from a power-on and therefore avoid any inadvertent triggering during the operation of the circuit.

The use of this type of active clamp has several advantages:

– as the protective component relies on the conduction of a MOS transistor, it is far less sensitive to fluctuations of the process than protections based on parasitic LNPN transistors;

– this allows for greater portability between the different manufacturing units for the same technology, which caused problems for LNPN-based protection systems;

– the ESD protection design shifts from technological optimization to circuit design;

– ESD protection strategy verification can be carried out using classical simulation tools, such as SPICE, used by circuit designers.

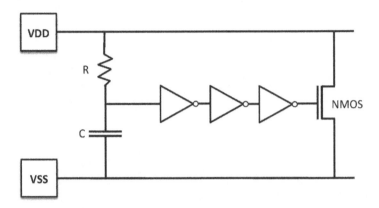

Figure 3.16. *ESD power supply protection or PC protection based on a large NMOS transistor triggered by a dynamic RC circuit*

The disadvantages include the following:

– the need for a large MOS transistor, which takes up a considerable silicon area: depending on the size of the chip, two or more clamps of this type, of the size of a bonding pad, are needed for the ESD current to be absorbed;

– the trigger circuit also takes up a non-negligible amount of space due to the need to generate a delay in the order of 100–1,000 ns. It can take up 20–40% of the total surface of the global protection system;

– a major disadvantage involves the protection of large chips as this robustness is then greatly affected by the resistance of the metallic VDD and VSS rails, meaning that additional layout precautions must be taken for the metallic rails and for the positioning of the required clamps.

An approach proposed by Motorola [TOR 02] to ensure uniform robustness for all of the inputs/outputs of a large integrated circuit involves establishing a distributed ESD protection network. In this way, instead of a small, limited number of big clamps, a small active clamp is placed at each input/output pad. The challenge of this approach is that the constraint of the silicon surface area occupied by the size of the trigger circuit cannot be reduced as it is linked to the RC time constant to be generated. The solution proposed by the authors is to use an equally distributed trigger circuit with a large RC network located in the power supply pad and small complementary capacitances distributed in each I/O.

As the cost of silicon increases with advanced technological nodes, it is important to keep the surface allocated to ESD protections to a minimum. In this vein, Stockinger *et al.* [STO 05] have proposed another improvement for the distributed active clamp by implementing an additional rail called Boost, which allows the voltage V_{GS} applied to the gate of the big MOS clamp to be increased, and therefore to improve its transconductance. This approach allows the size of the MOS to be divided by a factor of 2.3, reducing its silicon footprint.

3.2.5.1. Thyristors

The thyristor, or SCR (Silicon Controlled Rectifier), is another bipolar component that is increasingly used as an ESD protection structure, especially in advanced CMOS technology. Compared to the MOS transistor, and even to the bipolar transistor, it boasts greater ESD efficiency in terms of on-resistance, robustness, leak current, parasitic capacitance and therefore silicon footprint.

Three weak points have limited its use for a long time:

– a high trigger voltage;

– a high trigger time resulting from the latching effect between the two parasitic bipolar transistors NPN and PNP;

– a very low snapback voltage (~1.2 V).

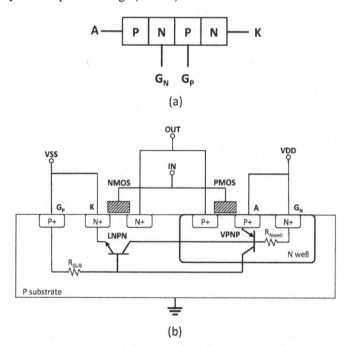

Figure 3.17. *SCR base structure (a) and illustration of the parasitic SCR in CMOS technology (b)*

The SCR is made up of two looped bipolar transistors, PNP and NPN, the whole constituting a base PNPN structure, as presented in Figure 3.17(a). This structure comprises three junctions in series (PN, NP, PN), connected such that the central junction is inverted. The electrodes at the extremities, A and K, are called Anode and Cathode of the SCR, respectively. The electrodes G_N and G_P are called gate N and gate P. There are therefore different biasing modes of the SCR, defined by the different possibilities for connecting the two gates. If no connection is established between the gates, the dipole structure is called a Shockley diode. Normally, all the gates are connected in order to reduce the risk of triggering the SCR, called a latch-up. In CMOS technology (Figure 3.17(b)), gate G_N and the anode are short-circuited and connected to VDD. This is also the case for gate G_P and the cathode, which are both connected to the ground. In this configuration, the SCR is in a blocked mode. To activate it, the voltage at the terminals of the device must become greater than the breakdown voltage of the central junction. This breakdown voltage is usually greater than 10 V in advanced CMOS technology. This operating mode therefore cannot be used. The trigger voltage can be reduced by connecting gate G_N and/or gate G_P to an external trigger circuit, which must be able to take gate G_N to a low potential, or gate G_P to a high one.

The first method used to decrease the trigger voltage of the SCR involved a highly doped diffusion P^+ in the middle of the well/substrate junction, which means decreasing the value of the breakdown voltage to that of the P+/N-Well junction. This breakdown voltage remains high, however, and the best solution involves adding a trigger circuit. The simplest method is the LVTSCR [KER 99a], which involves integrating an MOS component configured in GGNMOS across the P+/N-Well junction (Figure 3.18(a)). A PMOS component can also be integrated. Chen and Ker [CHE 07] proposed this solution coupled with an RC trigger circuit, allowing the trigger voltage of the SCR to be decreased further (Figure 3.18(b)). The disadvantage of integrating the trigger circuit into the SCR is that it makes the structure bigger, and therefore introduces a greater transit time of the carriers for the clamping when in SCR mode. An external trigger circuit is therefore more appropriate. A final solution, the advantage of which lies in the possibility of adjusting the trigger voltage, is to use a chain of diodes for the trigger (Figure 3.18(c)) or DTSCR [DI 07]. Its main disadvantage lies in its high level of leakage current, which increases with the operating temperature.

The SCR has excellent ESD robustness, with a small silicon footprint, but its disadvantage is its snapback characteristic, with a holding voltage V_H in the order of 1.2 V (Figure 3.19). Its use was therefore limited up to node CMOS 32 nm, whose supply voltage VDD was reduced to 1 V, thus allowing it to be used without fear of triggering the latch-up, as it was lower than V_H. This also allows it to be considered as a centralized protection, and thus to considerably reduce the surface area dedicated to the ESD protection.

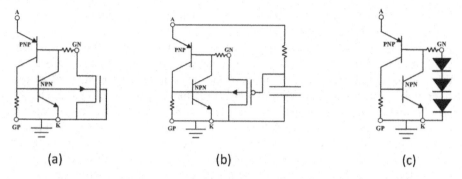

(a)   (b)   (c)

Figure 3.18. *ESD protection based on an SCR triggered using a GGNMOS (a), on a PMOS triggered by an RC circuit (b) and on a chain of diodes (c)*

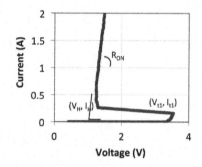

Figure 3.19. *TLP characteristic of a SCR-based ESD protection [BOU 11]*

STMicroelectronics conducted several studies on the optimization of ESD protection systems based on SCR for its advanced CMOS technology [GAL 12, BOU 11]. One original approach is focused on the optimization of a global and distributed protection strategy, relying on a network of several triacs [GAL 13], called Beta-Matrix. In this network, each triac is made up of two thyristors, as shown in Figure 3.20(a). The use of triacs, which are bi-directional components, results in perfect symmetry for each ESD stress bias. The physical implementation, which is shown schematically in Figure 3.20(b), involves creating a matrix of thyristors in a single N well (NWELL in the figure), each individually housed in a P

Well (PWELL in the figure). Thanks to the diagonal metallic interconnection shown in the matrix (VDD, VSS and I/O), two by two the thyristors form several triacs. Each P Well is either the anode of several triacs or the cathode of other triacs. A single gate G_N formed by establishment of contact N^+ on the N Well controls all of the triacs of the network. The trigger network linked to the gate is made of an RC filter and an amplifier (made with two inverters) for the positive ESD stress biases and a simple diode for the negative ones. A Beta-Matrix is in this way set up in each of the input/output (I/O) pads, and there is no need for a large PC at the height of the power supply pads, since it is distributed around each I/O. This strategy has been validated in CMOS 32 nm technology and allows for an ESD robustness of 2 kV HBM, 100 V MM ND 250 V CDM with perfect immunity to latch-up at 25 and 125 °C.

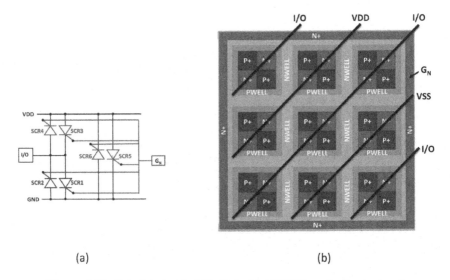

(a)   (b)

Figure 3.20. *Principle behind the Beta-Matrix ESD protection network [GAL 09, BOU 11]: electrical schematic (a) and simplified overview of the layout implementation (b). For a color version of this figure, see www.iste.co.uk/bafleur/esd.zip*

For more high voltage technologies, such as smart power technologies, the use of an SCR is trickier, and it must be optimized in order to ensure immunity to latch-up during circuit operation. An original approach proposed by the LAAS-CNRS [ARB 12] consists of combining an MOS transistor, an IGBT and an SCR into the same component, as illustrated in Figure 3.21(a). The goal is to replace the MOS PC by this component in order to reduce the silicon surface and increase ESD robustness performance. The initial component is a LDMOS power component into which an IGBT is inserted by implementing P^+ regions in the drain, as can be seen in the 3D view of this mixed component in Figure 3.21(b). Control of the holding voltage V_H and current I_H is optimized by the ratio of the P$^+$ regions in the drain, the reduction of the channel (up to 30%) by introducing a P$^+$ region in the source diffusion, in the shape factor of the component, as well as the length of the drift zone of the LDMOS [ARB 15]. The holding voltage V_H can thus be increased to a value greater than 5 V and make the structure usable in the ESD protection of the low voltage logic of the technology. For higher voltages, several structures can be put together in series to increase the holding voltage. The ESD robustness is particularly high with an ESD current of 8 A, equivalent to 12 kV HBM, with a small silicon footprint (1.5 mA/μm^2, outside of the trigger circuit).

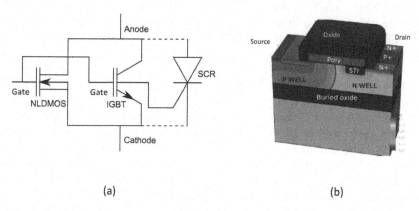

(a)   (b)

Figure 3.21. *Mixed ESD protection, with an MOS, IGBT and SCR in the same component: electrical diagram (a) and 3D view of implementation onto an SOI technology (b). For a color version of this figure, see www.iste.co.uk/bafleur/esd.zip*

3.2.5.2. *ESD protection of mixed technology*

The particularity of mixed technologies is that multiple, low, mid and high voltage power supplies must coexist within the same chip. These different power supply voltages require specific precautions to be taken for ESD protection strategies.

The first clear difficulty is that one single type of protection structure will not be enough to protect all of the inputs/outputs. Usually, a specific design is needed for each type of pad to be protected, depending on the range of associated voltages.

The second issue lies in the implementation of equivalent discharge pathways for all of the combinations of ESD stress.

Finally, a third issue is the possibility of accidentally triggering the ESD protection through noise from the digital blocks. This problem can be resolved by separating and isolating the analog and digital power supplies. The classical technique involves implementing bi-directional diodes (two back-to-back diodes) between the different power supplies, and several diodes in series can improve immunity to noise.

To illustrate this, in Figure 3.22, we provide an example of a global protection strategy of a mixed technology based on a central ESD1 protection and local mono-directional protections between each couple of pads (ESD2–ESD9). For this simple case, let us look at the various discharge paths. A positive stress on I/O$_1$ relative to V$_{DD1}$ involves a simple and direct path with the protection ESD2. On the other hand, a negative stress on these same pads results in a much longer discharge path involving ESD6, ESD1, ESD8 and ESD3. The worst case in terms of length of the discharge path is shown in this figure. This is the stress between the two I/O$_1$ and I/O$_2$ pads, which, for a positive stress, involves ESD2, ESD6, ESD1, ESD9 and ESD5, and for a negative stress, ESD4, ESD7, ESD1, ESD8 and ESD3.

This simple example shows that the approach is not well adapted to guaranteeing an equivalent ESD robustness over all of the input/output pads, and that for some discharge pathways, in addition to the on-resistance of each protection, there is also the impact of the resistance of the power supply rails. As a result of these different resistances, there is a high risk of overvoltage in the internal circuit and therefore of failure caused by an ESD stress. This problem would not exist if the ESD protections were bi-directional as in the Beta-Matrix approach described in section 3.2.2.2. Such a bi-directional approach has also been proposed by Albert Wang [WAN 06]. For bi-directional protections between VSS rails, or between different VDD power supply buses, the simplest typical approach is to use back-to-back diodes: this is two diodes for VSS (when they are all at 0V) or several diodes set up in series for the VDD rail.

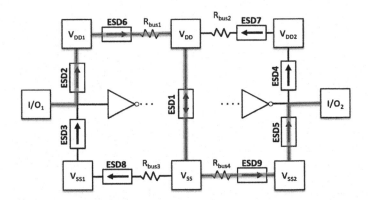

Figure 3.22. *Example of a global protection strategy of a multi-supply mixed technology, where V$_{DD}$>V$_{DD1}$>V$_{DD2}$. The discharge path is shown in orange for a positive stress on I/O$_1$ relative to I/O$_2$. For a color version of this figure, see www.iste.co.uk/bafleur/esd.zip*

In order to reduce the risk of failure in the internal circuit during a complex discharge, one approach proposed is to set up ESD rails where these discharge currents can circulate [KER 99b], following the same principle as the centralized protection approach. The ESD discharge is carried from the input/output (I/O) pads via a mono-directional protection (diode) toward the power supply rails. Each power supply rail is then connected to a specific ESD rail via a bi-directional protection. Depending on the size of the block, this bi-directional protection can be implemented in a distributed manner. Finally, the discharge is absorbed by the central protection, placed between each of the ESD rails. An example for three power supplies (VDD1, VDD2 and VDD3) is provided in Figure 3.23. For this bi-directional protection, Ker *et al.* propose thyristors controlled by an MOS. In this example, there is therefore a rail linked to each power supply (ESD-BUS_1, ESD-BUS_2 and ESD-BUS_3) and a single rail ESD-BUS_0 that connects the different VSS of each block, as each is at 0V in this case. Finally, there is a central protection between each couple of the ESD rails, written ESD CLAMP. For this clamp, the authors use cascoded SCR devices.

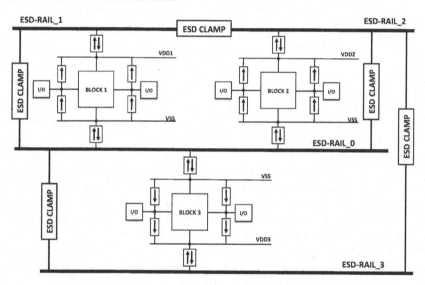

Figure 3.23. *Example of a global protection strategy of a mixed technology based on the use of an ESD bus for a chip with three different power supplies. In this example, VDD1 < VDD2 < VDD3*

3.3. Discrete protections

For various reasons (cost, technology capabilities and performance levels), it is not always possible to provide an integrated circuit with the ESD robustness needed to survive in the real world. When this circuit is implemented onto an electronic board, one solution involves adding an external protection, such as a TVS (Transient Voltage Suppressor) [JIA 11], for example. Different types of TVS, able to deal with the stresses at the system level, are connected on the electronic board at the input/output ports [MAR 09], as illustrated in Figure 3.24. The goal of this protection is to limit overvoltages and overcurrents caused by the ESD on the input or output ports of the circuit being protected.

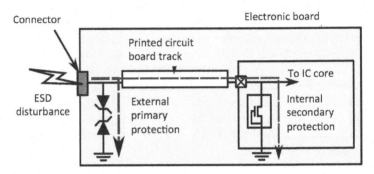

Figure 3.24. *Addition of an external protection on an electronic board to protect the system from discharges occurring on one of the connectors*

However, the fact that an external protection handles the stresses at the system level does not mean that it will automatically protect the pin of the circuit when it is connected in parallel to this circuit. The characteristics of the external protection must be considered, as must the integrated protection circuit, as well as the impedance of the track of the printed circuit, or there is a risk of bringing no additional protection, or even degrading the existing one. The role of the external protection, or primary protection (Figure 3.24), is to re-direct the strong currents toward the ground. However, as the system stress levels are very high and TVS-type external protections have a high on-resistance, the voltage induced at their terminals is very high. The secondary protection, Figure 3.24, is the protection that exists inside the integrated circuit. Its role is to ensure that the circuit pin can absorb a certain quantity of current without being damaged. The robustness levels of these protections are determined by the HBM, CDM and MM tests. The impedance between the primary and secondary protection plays a critical role. The voltage induced by the primary protection being high, the line impedance must limit the current in the direction of the circuit to a value that is compatible with its intrinsic robustness. Take the example of a circuit whose robustness is 1 kV HBM. The pulse current that the circuit can sustain is 1,000 V/1,500 Ω = 0.67 A. If the voltage induced by the TVS during the stress is 100 V, the line impedance must be greater than 150 Ω in order to limit the current to 0.67 A in the circuit pin (100 V/150 Ω = 0.67 A) [MAR 09].

Furthermore, the trigger voltages of the added external protection need to be verified relative to the internal protections. The strong current characteristics of an external protection superimposed onto those of an internal protection are represented in Figure 3.25 for two cases, (a) and (b). The internal protection of the circuit in case (a) has a lower robustness than in case (b) (see the level I_{t2}). However, in case (b), the external protection cannot protect the circuit as its trigger voltage is higher than the voltage of the circuit. Most of the current flows through the internal protection of the integrated circuit, which is not designed to deal with the high energy levels of a system stress. Contrarily, in case (a), the internal protection of the circuit has a greater trigger voltage, and the circuit can be protected efficiently by the external protection. This sinks most of the discharge current [STA 09].

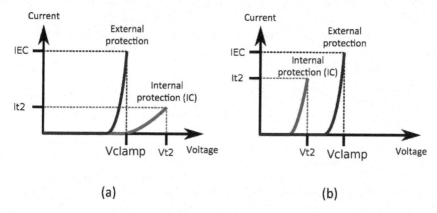

(a)   (b)

Figure 3.25. *Strong current characteristics of an external protection superimposed over an internal protection [STA 09]. Case (a): internal protection with a trigger voltage greater than the discrete protection; case (b): internal protection with a trigger voltage lower than the discrete protection*

3.4. Challenges of the protection strategy at the system level

As mentioned at the start of this work, the risk of an integrated circuit becoming damaged by an ESD is not limited to the phases of manufacturing and assembly. Discharges can take place in the final electronic product, and propagate throughout the inside and cause failures. Integrated circuits are usually soldered onto a board and assembled in a package for the application. As a first approximation, we can consider that the circuits are protected if they have successfully passed the HBM, MM and CDM robustness tests. Unfortunately, this is not always true. The system operates in the end usage environment, meaning that the discharges are no longer limited and are more powerful. Thus, even if the circuit is robust relative to HBM, CDM and MM events, this does not guarantee that it will survive in the system [HYA 02].

The notion of the system varies greatly depending on the speaker, and as such we would like to note that here we are only looking at the electronic board or PCB (Printed Circuit Board). In electronic equipment, integrated circuits are at the heart of the application, and are therefore also the main cause of failure, although the poor robustness of some passive components must not be neglected.

Two types of failure must be considered in terms of the system. The first concerns material robustness (*hard failure*), which leads to the destruction of the component or of part of its functionality. ESD can also cause transient failures leading to system dysfunctions. In this case, we talk of susceptibility to ESD or *functional failure* (or *soft failure*).

As illustrated in Figure 3.26, it is hard to predict how an ESD event can propagate from outside of the system (here an electronic board) into the component.

When the stress arrives through a connector and propagates along the board through various elements up to the integrated circuit, it is referred to as propagation by conduction. In the worst case, the disturbance can reach the component if one of its pins is directly connected to the outside. This is the case of pins used in communication protocols, like Ethernet pins, or "LIN" (Local Interconnect Network) or CAN (Controlled Area Network), a circuit commonly used in vehicles. External protections like diodes or TVS can be added onto these pins on the discharge path to protect the circuit. The sizing of an external protection like this is presented in detail in section 3.3, as long as the information regarding the protection techniques used by the circuit manufacturer is available. Similarly, depending on the system configuration, the external protection can create a new discharge path for the current toward a less protected pin of the component. It can also change the discharge path inside the component in relation to the initial protection strategy.

The disturbance can also propagate indirectly, by electromagnetic radiation. In both cases, the propagation phenomena can be complex. The coupling mode depends on the geometrical configuration, the routing of the electronic board, and on the board's component. In this case, it is hard to predict the waveform of the ESD stress at the component input, and therefore to deduce the behavior of the circuit or to predict the failure.

Whichever the coupling mode, conducted or radiated, it quickly becomes clear that the propagation of the stresses through the systems requires specific analysis.

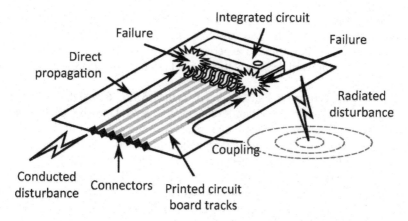

Figure 3.26. *Illustration of the disturbance by conduction or radiation inducing failures in an electronic board*

As mentioned previously, the manufacturers of integrated circuits have been constantly developing a large number of internal protections in chips, to increase their reliability during ESD events. In an integrated circuit, interactions between protections are common, and to this day, no model can predict their behavior in the final system, whether it is powered or not. There are several main reasons for this:

– The lack of releasable information on the protection networks. As reliability is strongly linked to the robustness of components, circuit manufacturers tend to keep their know-how secret. Providing the protection strategy to the equipment manufacturers (for whom the component is a black box, literally) is therefore not a possibility.

– The complexity of the interactions between the circuit and the rest of the system. Electronic boards integrate very large numbers of components and must operate in tough environments. The real system is made up of a large number of elements, with varying levels of impedance: connectors, tracks of the printed circuit, passive elements (decoupling capacity), circuit package, etc. When the discharge propagates throughout the system, each impedance loss changes the aspect of the signal. There are also the changes in impedance of the ESD protections that are external or internal to the circuits, which, ideally, go from infinite (open circuit), to zero (closed circuit), when the protections are triggered. The switching from the blocked state to the passing state can be nonlinear and complex (case of the SCR), with variable time constants. It is therefore very hard to predict the non-controlled reflections that propagate and become superimposed over the transmitted waveforms in the case of rapid transients (1 ns).

– The multiple interactions between all of these elements are still poorly controlled, and given that the issue of reliability of systems against ESD is recent, designers in this field have limited experience. Moreover, there is no design tool that is entirely oriented toward dealing with powerful rapid transient stresses.

As shown in section 1.2.1 in Chapter 1, standard IEC 61000-4-2, which is the main standard aimed at ESD system level tests, has several drawbacks, especially in terms of the reproducibility of tests, and issues with the radiation of the guns. As a result, it is hard to interpret the results when electromagnetic disturbances, which are not reproducible between testers, become coupled to the system. As ESDs are pulse disturbances, the spectral distribution can go up to several gigahertz [JIU 01]. Quick transition times, in the order of a few hundred picoseconds to several dozen nanoseconds, linked to high current/voltage amplitudes, result in large frequency bandwidths. The strong dI/dt and dV/dt transitions traveling through the tracks of the boards and through the cables induce electromagnetic radiation that becomes coupled to the various elements and circuits that make up the system.

Considering these effects, determination of the current and voltage waveforms at a given point in the system is a big challenge. For example, when the destruction of an integrated circuit occurs during a system test, designers must ask themselves the following questions:

– Which current is reaching the circuit when a discharge is injected into the connector?

– How can we understand the mechanism behind the destruction of the circuit despite there being no information available regarding it?

Equipment manufacturers are quickly confronted with the lack of knowledge of measurement techniques and of information for the analysis of the injection point and of the propagation of the stress in the system.

To solve these issues, system designers must therefore turn to the integrated circuit manufacturers to ask them for additional information regarding the circuit. However, they do not provide any additional information for reasons of intellectual property, which in most cases leads the equipment manufacturers to requiring a specific level of robustness of the component, most of the time in line with the JEDEC/ESDA HBM standards. As system ESD is the most severe [STA 09], the HBM level required for the circuit, which is usually 1 kV, is increased to a higher value. This increase, however, does not guarantee better robustness of the circuit in its final application. The system's protection must be carried out by all of the elements that make it up and not only by the integrated circuit.

The model of the integrated circuit is at the heart of the issue as it currently presents most of the system functions, whether they be analog or digital. As shown in Figure 3.27, in which the component is placed at the center, the complex interactions between the system (passive–active–tracks) and the component must be taken into account. Each of these elements can be complex, and an analog simulation (of the SPICE type) cannot be considered as the simulation time does not allow it (this point will be covered in Chapter 4 when comparing behavioral models and classical analog models). As component manufacturers do not provide information on the protection strategies of their circuits, the equivalent impedance of the circuit cannot be known (in high injection states), nor can the current path throughout the component be known. As a summary, it is mainly the different interactions between the elements of the system, regardless of the size, that are difficult to consider when looking to predict the impact of ESD.

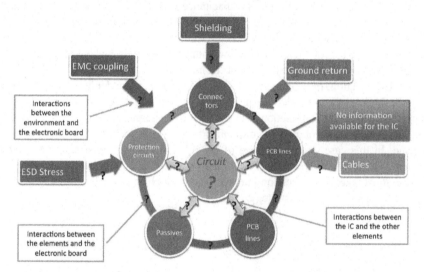

Figure 3.27. *Interactions between the various elements of a system. For a color version of this figure, see www.iste.co.uk/bafleur/esd.zip*

There is a clear need not only for models, and modeling tools and prediction tools that take into account both the component and its protections, but also for all of the other elements that are included in electronic boards, in order to anticipate failures. These methods and techniques are vital for understanding the various phenomena caused by conduction or by radiation.

Currently, there are simulation tools that could be used for this, such as electromagnetic simulation software, and/or high frequency electrical software. Their use, however, is complex, and they require long calculation times. Faced with the complexity of the systems to be analyzed, their use cannot be considered, due to the number of active circuits that would have to be considered in one system. On the other hand, as stated, no ESD information is provided regarding integrated circuits, and the issue is therefore on how to supply the software. Currently, simulations of the system undergoing ESD are, as a result, difficult or even impossible.

Knowledge of the phenomena that govern the distribution of the ESD currents throughout the board is the first step. However, the issue of ESD at the system level does not stop there. As illustrated in Figure 3.28 [MON 10, MON 11a], from the point of view of the prediction of the dynamics of the currents circulating through the integrated protections, it is vital that high frequency (HF) elements be taken into account for passive components. The diagram shows the simple case of a capacitance in parallel on the input of an IC whose internal protection is a snapback protection. In the case of a TLP stress, the external capacitance is charged using an ESD current, up to reaching to a value of the trigger voltage of the protection structure (charge of a capacitance at a constant current). When the value of the trigger voltage is reached, the protection characteristic breaks over a low voltage value, thus causing the discharge of the capacitance in the protection. In the example presented in Figure 3.28, during a 100V TLP injection, the voltage climbs up to 70 V before suddenly going back down to 17 V. This injection level corresponds to a current of 1.4 A, measured without the capacitance, which in conjunction with capacitance generates a peak of more than 6 A. The current that the component must absorb is equal to nearly four times the one injected. In this case, the importance of the relation that can exist between external elements is clear, as is the protection strategy developed within the component. This strategy must take the application into account as much as possible so as to determine the most adapted protection strategy.

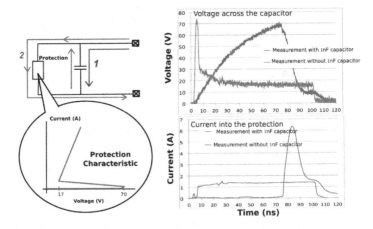

Figure 3.28. *Interactions between the various elements of a system. For a color version of this figure, see www.iste.co.uk/bafleur/esd.zip*

3.5. Conclusion

The protection of electronic systems with regard to ESD continues to be a challenge for system designers due to, on the one hand, the rapid evolution of integrated circuit technology and of electronic boards and, on the other hand, the large-scale introduction of these systems in a number of applications involving the safety of individuals (cars, aeronautics). On top of this, there is a strong constraint relating to the cost of the protection system, which, when they are integrated into advanced technology, must have as small a silicon footprint as possible.

The considerable limitations, as much in terms of performance levels as in cost, provide a major motivation for the establishment of modeling methods and simulation methods allowing these protection strategies to be optimized from the design of the integrated circuit or of the board, thus minimizing the number of design iterations.

In Chapter 4, we shall look at the various modeling approaches put in place for the optimization of the ESD protection strategy, from the component to the board system.

4

Modeling and Simulation Methods

The efficiency of a protection strategy is largely dependent on the design approach implemented. For a long time, this approach was based on the experience of the designer and an empirical approach of trial and error could have a heavy impact on the price and the time to market of the product. The rise of computer-aided design tools and their simulation methods provides an approach that is more predictive of the ESD robustness of an integrated circuit and limits the number of design iterations associated with ESD protection.

In this chapter, we describe the various modeling and simulation methodologies used to optimize the ESD protection strategy. For the protection component, the physical simulation or the TCAD (Technology Computer Aided Design) allows for an in-depth understanding of its behavior under high currents and strong electrical fields. It can even predict the robustness of the protection structures, depending on the geometry and technological options. For an integrated circuit, a traditional approach is to use the same electrical simulator that was used for the circuit design, which requires compact modeling of the protection components. For a printed circuit board system, we shall present the predictive modeling approach developed at LAAS-CNRS, which is currently being considered for a transfer to ESDA standards.

4.1. Physical simulation: TCAD approach to the optimization of elementary protections

For a long time, ESD protection structure design was carried out in an empirical manner, using standard components of the various technologies (such as diodes, bipolar transistors or MOS) and applying rules-of-thumb that are specific to ESD (ballast resistance, width of metallization, etc.). The size was optimized by iterations between manufacturing and destructive tests depending on the level of robustness required. Simulation tools allow the ESD problem to be taken into account from the development stage of a new technology and predict the efficiency of the proposed protections.

Physical simulation tools are used to assess the behavior of a component by using its two-dimensional description. For simple cases, and when it is truly necessary, a three-dimensional simulation can be carried out but with a far longer computation time, as well as with a more limited mesh. However, the overall simulation of an integrated circuit is currently impossible with these tools and must be reserved for SPICE electrical simulators.

The use of physical simulation tools began in university circles [PIN 84, NOL 94]. Since the beginning of the 1990s, commercial tools appeared such as S-PICES (Silvaco), MEDICI(2D) and DAVINCI(3D) (TMA, Avanti and then Synopsys), and DESSIS (ISE then Synopsys) [SIL 16, SEN 16]. These tools offer a wide variety of models, are generally three-dimensional and also include materials other than silicon, such as III-V or SiGe.

No matter the version of the tool used, whether university or commercial, these physical simulators are based on the same principle. The structure is made using a two- or three-dimensional mesh. The critical information needed is the geometry, the composition of the different regions, the doping of semiconducting regions and the position of the electrodes. The semiconductor equations, which can be expressed through differential equations, become discrete on this mesh and are resolved by using an adapted numerical method. Voltage or current conditions are fixed on the electrodes and the solutions are found through iterations once the convergence criterion is met. The user has a choice between several numerical methods and can adapt the parameters to the simulated component. Several operating modes are available, such as a static or quasi-static mode, a transient mode, or even a small-signal sine wave mode. It is the transient mode that is used for ESD, though it often corresponds to the longest simulation time.

The solution to the heat equation for the crystal can also be used. The diffusion of temperature in the silicon is not actually critical in relation to the duration of the electrostatic discharge. Indeed, the thermal diffusion constant in the silicon is in microseconds for dimensions in the range of a micrometer; however, the time needed to reach maximum current during an ESD discharge is closer to 10 ns (in the case of HBM), if not 1 ns (in the case of CDM). The heating of the chip is thus highly localized. This local increase in temperature has a significant effect on the physical parameters of the models used. We shall later show that this is problematic in terms of their validity.

There is a large choice of models, the selection and main parameters of which are left to the discretion of the user, depending on the simulated component. The model equations are usually not modifiable but for some there are external modules in which the model can be rewritten in a more advanced language. The new model is then interpreted, which can be extremely costly in terms of simulation time, or compiled and linked to the simulation core, which is a more efficient solution.

Two types of results are obtained after the simulation. Either the current/voltage characteristics on each electrode of the component can be compared to the experimental measurements or internal cross-sections are used to give the distribution of the different physical values in the component. These distributions are not usually measurable directly in the experiment but in certain cases they can be compared to failure analyses (heated areas in the volume) or to optical measurements (see section 2.2 of Chapter 2).

4.1.1. Description of the structure

There are two methods used to define the component to be simulated: an analytical description or a technological process simulation.

The first method involves defining the geometry of the structure by using line segments to define the doping zones through analytical equations and to generate meshing. These equation parameters are chosen to fit the values of the doping profiles obtained through a SIMS (Secondary Ion Mass Spectrometry) analysis. It is also possible to deploy the SIMS profiles directly onto the structure and the doping values are then interpolated onto the meshing. This option is only applicable in the vertical direction.

The second solution uses the initial simulation of the technological process of manufacturing. For this, access to a technological simulator is necessary as is knowledge of the different technological steps and parameters necessary to describe them. In the case of recent technological procedures, the number of technological steps is significant (>20 mask levels), making these simulations very heavy. However, when they are well-calibrated they accurately show the geometry and the doping in the structure. The point of this method is to obtain a meshing upon which lateral doping profiles are more realistic than in the analytical approach. Indeed, in the latter, vertical profiles, whether obtained analytically or by reading a SIMS profile, are shown in the lateral dimension by using an extension factor (usually chosen empirically between 0.3 and 0.8). Figure 4.1(a) shows an example of a measured SIMS profile, which is the same profile generated by the analytical description tool and the simulation results of the technological process. For vertical profiles there is a good concordance between the three curves. Figure 4.1(b) shows another example of the lateral profile of boron and phosphorus by using the methods described.

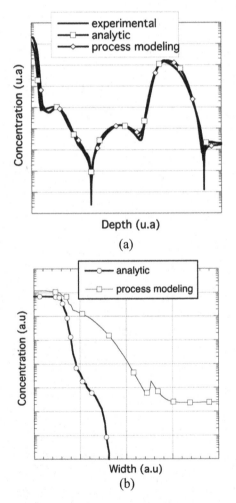

Figure 4.1. *Differences in vertical profiles (a) and lateral profiles (b) between the technological process simulation and the analytical description*

The diffusion of phosphorus defines the position of the junction and its graduality. There is a significant difference between the two profiles, the lateral extension here being 0.8. This has an important effect when the component is based on a lateral trigger process. Differences of more than 20% were found in the trigger voltage of the bipolar structures [SAL 05a].

The results of the simulation of the technological process are not immediately usable. Indeed, the number of points generated is high and the high point densities are often localized in regions that are of little interest to the physical simulation, such as the detailed description of oxide or metallization zones. It is common, however, for strategic areas (junctions, MOS channels, drift zones) to be described by a loose mesh. It is then necessary to optimize the meshing with a tool that is generally also used for the analytical description. We can use this stage to artificially add dopings to the structure and to adopt a mixed description strategy. This method is particularly efficient when the component consists of technological steps that induce high simulation times but whose lateral description properties are not critical. This is the case with buried layers in bipolar structures, for example.

Finally, if the work carried out concerns the design of protection structures on unfixed or prospective technologies, only the technological process simulation approach can be used.

4.1.2. Simulator calibration

Physical simulators come with several models, many of which are redundant, and with several sets of parameters. For silicon, and for component sizes that are much larger than nanostructures (the latter requiring quantic simulations), the main models are based on the following physical values: the width of the band gap, electron affinity, the intrinsic carrier density, the partial ionization of the dopants, the mobility of the carriers and the generation/recombination mechanisms. These models are in relation to different parameters linked to the structure and its state: the concentration of dopants, the temperature of the crystal lattice, the internal electrical field or the density of free carriers.

The choice of model depends on the component studied. Generally, we differentiate between models used for an MOS component and a bipolar component. The mechanisms that govern their operation are different and each type must have its own appropriate set of models. For ESD protection structures, this is different. If a NMOS transistor is used in the grounded-gate configuration (GGNMOS), it will be the parasitic bipolar transistor that is activated, inhibiting the field effect. In the majority of cases, models are chosen to demonstrate bipolar behavior. Only protections using a MOS and a capacitive gate coupling (GCNMOS) are exceptions to this rule. In this configuration, the field effect is used to trigger the structure with the bipolar effect then participating in the conduction. The choice of models is then even more difficult. This type of configuration can also be found in MOS power components that are self-protected against ESD [BES 02].

If the selection of models is carried out depending on the type of component to be simulated, their parameters can vary according to the technology used. It is therefore necessary to carry out a parameter calibration phase.

Calibration is first carried out on standard components of the technology. In the case of a bipolar transistor, two static characteristics can be useful as they are in close relation with the dynamic parameters of the ESD protection. The first characteristic is the breakdown voltage BV_{CB} of the collector–base junction, which demonstrates the avalanche phenomenon. When the ESD protection is drawn correctly, the value of the trigger voltage of the structure V_{t1} becomes closer to that of the voltage BV_{CB}. If we wish to delay the trigger until higher voltages, it is possible to change the internal base resistance that is responsible for the conduction of the base–emitter junction and therefore of the bipolar transistor. In any event, the voltage BV_{CB} must be evaluated correctly in static. The model of generation by avalanche is unique:

$$G = \alpha_n \,.n.v_n + \alpha_p \,.p.v_p \qquad [4.1]$$

This is a function of the density of the free carriers n and p, their travel velocities vn and vp, and the two ionization coefficients α_n and α_p. For the latter, different expressions are available. Two models [OVE 70, LAC 91] are derived from Chynoweth law's model and another model is entirely empirical [OKU 75]. The range of temperatures of these models is relatively limited (from 300 to 400K). More recent work from the University of Bologna has resulted in a new model whose valid range of temperatures has been extended to 700K [VAL 99]. Table 4.1 compares the static breakdown voltage measured and simulated on a bipolar component. In this case, it is the Okuto/Crowell model that gives the best value but it is not to be generalized to each case, especially as the spread of the results is not large for this component (<6%).

	$BV_{CB}(V)$
Measurement	*42*
VanOverstaeten model	44
Lackner model	44.7
Okuto/Crowell model	43.8
Bologna model	44.6

Table 4.1. *Example of breakdown voltage simulation using different ionization coefficient models*

The second static characteristic that is useful in calibrating the simulator is the direct current gain of the bipolar transistor protection. This parameter has a direct effect on the holding voltage V_H of the protection structure when the bipolar transistor is triggered. This voltage can be approximated using the following analytical relation [MIL 57, JAN 93]:

$$V_H = \frac{BV_{CB}}{(1+\beta)^{1/m}} \qquad [4.2]$$

where β is the direct current gain and m a factor between 2 and 6.

As such, we can see the importance of properly estimating this gain, as the breakdown voltage BV_{CB} has already been calibrated. Several physical values have an effect on calculating the gain and each one usually has several models in the simulator.

There are many models designed for the mobility of the free carriers. The temperature of the crystal lattice is taken into account in order to show its excitation and therefore the reduction of the mobility of the carriers that have been exposed to various scattering mechanisms [LOM 88]. Degradation of mobility can equally appear through carrier scattering through impurities [MAS 83, ARO 84], which depends on doping levels, or even through carrier-carrier-scattering [CHO 72, FLE 57]. Finally, certain mobility models consider the effects of surface states or electrical fields but are more applicable to MOS conduction structures. Philips' unified model [KLA 92] takes into account a certain number of these degradations and is often quoted within the literature as being well-optimized for bipolar transistors. In our experience, we did not find a true concordance between measurements and simulations by adjusting the mobility models [SAL 05b], although the Philips' model was the closest to it. Striker [STR 00] highly recommends the use of this model by showing a correlation between the measurements of the layers of the sheet resistance in relation to the temperature (see Figure 4.2). There was a small range of temperatures used (Tmax=150°C) so we may question whether the correlation remains as accurate as for higher temperatures.

In the same way, variation in the effective intrinsic density can affect the calculation of the transistor gain. Several models exist in the ISE simulator but none of them provide perfect correlation with the measurement. The model that gets the closest is the Slotboom model [SLO 77].

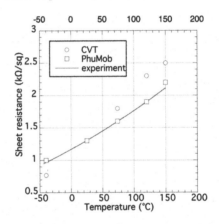

Figure 4.2. *Measurement and simulation of sheet resistance as a function of time [STR 00]*

Finally, it appears that it is the value of the carrier lifetime that results in the best adjustment of the simulated gain curve. This simulated gain, which enters the simulator as part of the calculation of recombinations, has an effect on the transport factor in the base α_T and therefore on the transistor gain. Analytically, this interaction is described by the following empirical equation [BAL 96]:

$$\alpha_T = \frac{1}{\cosh(W_B / L_n)} \text{ with } L_n = \sqrt{D_n . \tau_n} \tag{4.3}$$

with W_B being the width of the base, D_n being the diffusion coefficient and τ_n being the carrier lifetime.

Figure 4.3 provides an example of the calibration of the static gain of a bipolar transistor. The levels of the collector current can appear to have low values, much less than a milliampere, compared to the current levels I_H (a dozen mA) reached during a TLP stress. The pertinence of this calibration must therefore be brought into question. However, we cannot forget that the component is in self-biased mode and in a state of avalanche. The internal current is therefore strongly multiplied at the level of the collector–base junction in order to reach these current levels [TRE 04a]. This being said, the value of this gain is not that useful at higher current levels (several amperes), as in this high injection regime the value of the gain drops as the current increases [SZE 81]. However, as we shall see later, simulation of an ESD protection in this regime, where the temperature increases quickly, becomes questionable.

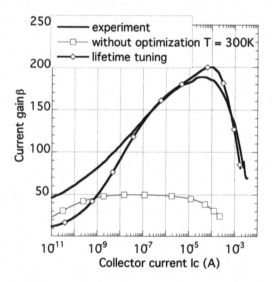

Figure 4.3. *Example of measurement and calibrated simulation of the current gain of a bipolar transistor*

4.1.3. Robustness prediction

In ESD robustness qualification testers, functionality tests are carried out between each stress in order to check the state of the circuit. The simplest, which is also the most commonly used, is to measure the leakage current of the circuit for a given voltage. This quick test often overlooks the appearance of latent defects and it is often more effective to measure the entirety of the characteristic. The failure criterion is a lot harder to find for the simulation. It is not possible to observe possible changes in the static characteristic after simulation of an ESD stress; the simulator cannot change the properties of the structure and reveal the appearance of degradation (fusion, dopant redistribution, filamentation, etc.). For this same reason, the cumulative effects of stress cannot be simulated. We must therefore observe the various physical parameters during an ESD stress on the structure and find a criterion over their values.

The most immediate criterion relates to the temperature of the crystal lattice. Several researchers [HON 93, GAL 02] are monitoring its maximal value and consider failure to have occurred when this value exceeds the temperature of fusion of silicon (1,685K). This criterion is highly debatable, as most of the models implanted in simulators are not validated at temperatures more than 600K. As the increase in temperature is caused by strong electro-thermal coupling, significant errors are likely using this method. However, the temperature can be used as a first approximation in order to optimize the protection structures. By identifying the hot spots of the volume, we are able to find solutions to reduce their values or to move them, which is usually reflected in an increase of the robustness with regards to ESD [TRE 02].

Other authors prefer to carry out TLP simulations in order to obtain the value of the failure current It_2. Indeed, it is sometimes possible during a simulation, at very high currents, to observe TLP curves in a second snapback[1], which can be interpreted as the destruction of the component [STR 00, ESM 01]. However, the level of the current found does not always correspond to the measurement. In other examples, the simulator diverges before this current level or no snapback takes place. In any case, the component undergoes very high current densities ($>10^5$ A/cm^2) and has hot spots with very high temperatures. The validity of the models can still be doubted and it is dangerous to qualitatively consider the electrical behavior of components in this configuration.

It was in an attempt to deal with this issue that we started our work on this theme, which is featured in the thesis by Christophe Salaméro [SAL 05b]. The result

---

1 The cited authors call this phenomenon "second breakdown". This is not very precise, and it would be wiser to call it a thermal breakdown or a second thermal breakdown. While the first breakdown (purely electrical) is not a source of doubt, and corresponds to the input of a component by avalanche in the blocked state, the second electrical breakdown is more ambiguous. In the case of bipolar transistors, we are used to referring to the second electrical breakdown as a high-current focalization point in the component. Given the high temperatures that exist in the focalized part, the carriers are generated thermally, thus reducing the amount of generation by avalanche and reducing the voltage that the structure is subjected to. The thermal runaway is irretrievable and results in thermal breakdown. The MOS community, however, considers the transistor to undergo a second breakdown when the electrical characteristic snaps back. In any case, the term "second breakdown" remains an ambiguous one in the literature.

of this work shows that the methods used provide good prediction results. It is based on low and medium current TLP simulations, within the validity temperature range of the models. During these simulations, the hot spots of the structure are identified and the values of the impact ionization rates, G_i, and of the recombination rates, R_{SRH}, are picked up at these points. G_i gives an image of the avalanche phenomenon, which is dominant in the structure until thermal instability is reached. From this regime, the effective intrinsic concentration of the carriers n_{ieff} becomes very high due to the increase of the temperature and becomes the source of the generation of carriers. As a result, the term $R_{SRH} \propto np - n_{ieff}^2$ is very negative. A polynomial law lets us extrapolate these values obtained at weak currents to higher levels; when both rates are found to be equal to each other we consider the point of thermal breakdown to have been reached. This method has been successfully applied in several protection structures, the robustness level predicted at 10%, and the location of the destruction correlated using failure analyses in a test structure. Moreover, this method has the advantage of being relatively quick, as bi-dimensional calculations are only carried out for low currents, ensuring a good convergence of the simulator. The simulations are usually twice as quick using this method. Figure 4.4 [SAL 05b] illustrates this method being applied to self-biased bipolar transistor ESD protection. The destruction current, predicted at 1.52 A, correlates exactly with the measured value of 1.5 A. The total simulation time ranges from 3 days for this method to 7 days for a prediction based on the detection of the second thermal breakdown, which also results in an error of more than 20% of the value obtained. This new method does present certain restrictions, however. It is applied to structures that are already optimized in terms of trigger homogeneity. It would be interesting to apply it to more complex structures and in the prediction of HBM robustness.

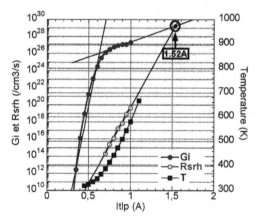

Figure 4.4. *Example of destruction current prediction of an ESD structure undergoing a TLP test. For a color version of this figure, see www.iste.co.uk/bafleur/esd.zip*

4.1.4. *Focalization phenomena*

The simulation tool makes use of a bi-dimensional description of the component. As the ESD waveforms are usually currents, the simulator transposes these values into current densities by using a factor that takes into consideration the depth of the component. This factor must be adjusted to the dimensions of the component to avoid overestimating these current densities. The cumulated width of the component is usually chosen but this means that the component only has one finger, or that the component is the same for all of the fingers, and that the current flows uniformly throughout the width of the finger.

4.1.4.1. *Focalization on a finger*

While in the static mode, all of the fingers of a multi-finger component can contribute similarly to the current flow. This is not the case in ESD structures and particularly those that exhibit snapback. During the rise of current caused by the electrostatic discharge it is possible, if the structure is badly optimized, for a single finger to be triggered. As the snapback voltage V_H is lower than the trigger voltage V_{TI}, all of the current flows through that finger, as the others are no longer in the avalanche state and the structure is not very robust. In order to limit this problem, protection structures have ballast resistors built into the collector and/or in the emitter which contribute to the successive triggering of the fingers, as illustrated in Figure 4.5.

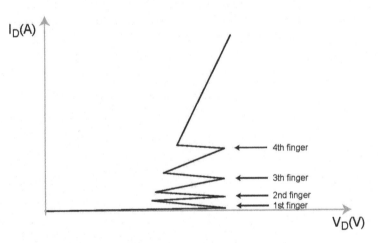

Figure 4.5. *Successive triggering of the four fingers of an ESD protection structure*

More complex simulations must be carried out in order to optimize these structures. The number of fingers is usually quite low (from 1 to 4) as the silicon footprint of the protection must be reduced. A first solution is to describe all of the fingers with the same mesh, assuming that they all have the same width. This solution presents the advantage of showing possible interactions between the fingers but the description of the structure is cumbersome and the number of points generated is high. Another configuration is preferred, which couples several fingers in an external electric circuit [NOL 02]. This mixed simulation, which is already used to describe the ESD pulse generator and its parasitic elements, allows us here to artificially add low value resistors (mΩ) onto the access to the various fingers. With no difference in value between these resistances, there is no reason for the fingers not to be triggered at the same time during the simulation. This resistance must not be confused with the resistance of the ballast, which is introduced by using a less-doped contact zone or by placing the metallization further away from the contact.

A compromise must be found for the ballast resistance: a value large enough to allow the triggering of the other fingers but not too big, so as to still ensure the protection voltage at high currents and provide low impedance for the structure. This method helps carry out this optimization.

4.1.4.2. Focalization on a zone of the finger

There are also focalization phenomena that exist over the width of the finger or the third dimension that the bi-dimensional simulator considers to be uniform.

When the component reaches the avalanche regime, the current flow initiates at the edges of the collector/base junction due to the presence of a spherical junction that experiences a larger electrical field. The avalanche current is spread over the entire width of the finger but the current densities remain higher around its edges. As a result, the bipolar structure is first triggered at that particular location. If no precautions are taken this localized high concentration of the current quickly leads to the destruction of the component but as the structures generally have a ballast resistance the problem is passed on to the high current levels [ESM 03, RUS 99]. As this resistance is distributed along the length of the finger, it limits the current at the critical area and allows for it to be spread along the entire finger.

Another phenomenon can intervene for much higher current levels. This is not really a focalization but rather a fluctuation of the density of the current that could create a hot spot over which the thermal breakdown would be initiated. Usually destruction is delayed, thanks to the beneficial counter-reaction of the temperature on the multiplication by avalanche. Fewer carriers are generated at the hot spot and the maximal current density is shifted toward a colder zone. A simulation over the width of the finger has been carried out to illustrate this phenomenon. To make the problem bi-dimensional several simplifications have been made, notably in terms of the base, which has been left as floating, and the results are considered qualitatively.

Figure 4.6 illustrates the rapid shifting of the current density peak for a vertical bipolar structure at the collector/base junction under the emitter [TRE 02].

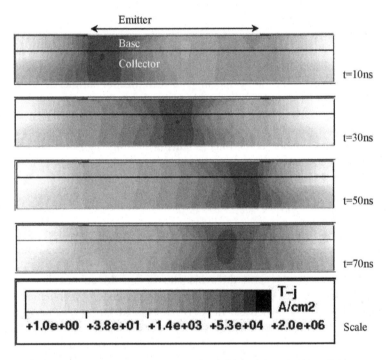

Figure 4.6. *Demonstration by simulation of the shifting of the value of the maximum current density along the finger of a self-biased bipolar transistor undergoing HBM stress*

This phenomenon has also been observed experimentally by Pogany using a laser interferometry bench [POG 03]. The time constants are also found to be very rapid.

There are not really any bi-dimensional solutions that can be used to highlight focalization phenomena along the width of the finger. The mixed approach used for multi-finger structures cannot be reasonably applied here. Indeed, protection structures typically have a finger length of around a hundred micrometers; if a first evaluation shows that the focalization takes place over 3–4 μm, this represents nearly 30 cells that must be placed in parallel. The description of such a structure is fastidious, as there are no schematic tools available to describe the electrical circuit. Moreover, the size of the problem makes the issue of simulation uncertain.

Problems of focalization along the length of the finger are less critical than the simulation of multi-finger structures. As the structure is dedicated to protection, ballast resistors limit these phenomena. An underestimation of the current density can take place at the triggering and can lead to an incorrect prediction of the holding voltage. This being said, the bi-dimensional simulations that we have carried out have never resulted in any large amount of errors for the estimation of this parameter. Fluctuations at higher current levels could be detrimental but the resulting error must be weighed against the fact that the structure is still triggered across the entire length of the finger, and that the validity of the models is questionable when close to high temperatures.

There remain cases where bi-dimensional simulation reaches its limits if we want to quantitatively evaluate the behavior of the structure relative to ESD. This is the case of the self-protected LDMOS power transistor for example, to which not all ESD rules can be applied, notably the introduction of the drain ballast which is incompatible with the need to have a low value of the on-resistance. We have seen that these structures, whose fingers are very long, display strong focalization phenomena that the simulation was not able to highlight.

4.1.5. Benefits of 3D modeling

As we have explained, there are focalization phenomena that take place during the ESD stress along the length of the active zone of the component but also across its width. The first effect is taken into account in a 2D simulation but a 3D simulation is strictly necessary in order to demonstrate the second effect. However, as this type of simulation is very complex to set up and requires limiting calculation times, the critical cases where it is necessary must be listed or at least the errors generated by a 2D simulation must be evaluated. If it becomes clear that for non-optimized ESD protection structures three-dimensional effects have a non-negligible impact, the case of well-ballasted ESD protections becomes less critical. Esmark [ESM 03] shows that for GGNMOS structures (CMOS technology 0.35 μm) and for a TLP current greater than 500 mA, there is no difference between a 2D and a 3D simulation. However, for lower current levels there are indeed differences. The errors can be especially significant for the snapback voltage V_H. The triggering of the structure takes place preferentially at the cylindrical junctions of the drain (or collector) [GAL 02]. Conduction located at this area increases the resistance of the structure, something that the 2D simulation cannot show. The 2D simulation sees conduction along the whole length of the finger and as a result underestimates the value of the voltage V_H. This case is not particularly vital for the end usage of the structure as the design margins are such that V_H must be greater than the power voltage of the circuit being protected in order to avoid latch-up issues. In this way, the 2D simulation brings a bigger constraint in terms of the design. However, the gap tends to get smaller as the ballast is increased.

However, in the case of low values of the ballast resistance, there are also high voltage differences between the 2D and the 3D simulations. The 2D simulation overestimates by 20–30% the level of the failure current I_{T2} during a TLP stress [ESM 01]. However, we must remain cautious in terms of the results of these simulations, as the failure criterion is often fixed as being the melting temperature of silicon and we have already expressed our reservations about the relevance of the results obtained at temperatures far greater than the validity limit of the models.

It therefore seems that the 3D simulation can provide an explanation for a specific point of the structure studied but its systematic use is not justified [VAS 03]. The bi-dimensional approach remains valid in most cases, especially if the structure being studied contains ESD design rules.

4.2. Electrical simulation: Compact modeling

Compact electrical modeling involves the use of known equations to accurately represent the behavior of the components, which generally provides a good convergence of the simulations. These equations, which are based on physics, are simplified as much as possible so as to have a minimum amount of parameters to extract. Strictly speaking, the parameters are not purely physical, and help adjust the model. The models proposed are therefore halfway between physical and behavioral. The most commonly used simulator for this type of modeling is the SPICE simulator [VLA 94].

The elements used as protection devices against ESD are active or parasitic components available in the technology (diodes, bipolar transistors, MOS transistors, thyristors). Their layout must normally be adapted for the high-current regimes and strong electric fields to which they are submitted under ESD stresses. In the same way, their electrical models must be extended to describe the physical phenomena that are characteristic of their function, such as avalanche breakdown, negative resistance characteristics, conductivity modulation, etc.

4.2.1. Diode modeling

The diodes are usually used in forward bias and it is therefore this mode that is modeled from the classical expression of the current:

$$i_d = i_s \cdot \left[exp\left(\frac{v_d - R_{ON} \cdot i_d}{N_f \cdot U_t} \right) \right] \qquad [4.4]$$

with i_s being the saturation current, U_t the thermal noise voltage, N_f the non-ideality factor, R_{ON} the value of the on-resistance, v_d the voltage at the terminals of the diode and i_d the current flowing through the diode.

The value of the saturation current i_s as well as the non-ideality factor N_f, allow the threshold voltage of the diode to be adjusted. In order to build the model, the three parameters (i_s, N_f and R_{ON}) are extracted by fitting them to measurement data.

For SOI (silicon on insulator), it can be beneficial to include the effects of self-heating into the model [WAN 00]. However, the complexity associated with this advanced form of modeling can cause issues in terms of the convergence of the simulator.

The reverse conduction of diodes is not modeled as this operating mode is not usually needed as part of the protection strategy of a circuit and robustness in this mode is very low. Only the capacitance of the diode is represented in the model in order to show the dynamic behavior of the diode. As a first approximation, the variation of the capacitance with the voltage can be neglected.

However, under certain configurations of the protection circuits, and when available as part of the technology, Zener diodes can be used as voltage limiters to protect the gate oxides. In this case, they are often associated with a primary stage that absorbs the strong discharge current. They can also be used as the trigger circuit of a protection (GGNMOS, thyristor or bipolar transistor).

4.2.2. Modeling of the bipolar transistor

The bipolar transistor is undoubtedly the most important component in terms of modeling of the ESD protection. Indeed, the issue of ESD first became apparent in

MOS technologies. However, MOS transistors are not usually designed to absorb strong currents under strong electric fields. Its NPN parasitic lateral bipolar transistor is therefore used to ensure ESD protection.

As seen in Chapter 3, this bipolar transistor is set up in the self-biased mode or triggered using an external source of current. When a negative discharge is applied to the collector relative to the emitter, the transistor behaves like a forward-biased diode. In the presence of a positive discharge applied to the collector relative to the emitter, in self-biased mode, the breakdown voltage must first be reached by avalanche of the collector/base junction in order to bias the transistor. This mode is not usually represented in models of the bipolar transistor, apart from the VBIC models [MCA 96]. However, these models take into account low rates of multiplication by avalanche which is not appropriate for ESD modeling.

4.2.2.1. Static characteristic

Let us recall the typical characteristic of a NPN bipolar transistor, as shown in Figure 4.7. Region I of this characteristic corresponds to the normal operation with a linear regime and a saturated regime. These two regimes are well-described by the standard models of the bipolar transistor. The two other regions correspond to behavior under a strong electric field (region II) and with a strong current (region III), which dominate during an electrostatic discharge. These two regimes are not considered in the standard model.

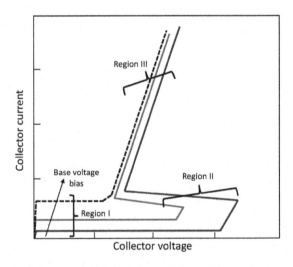

Figure 4.7. *Schematic electrical characteristic of a bipolar transistor for different base biases. Region I: linear and saturated regimes; region II: breakdown of the base–collector junction and snapback of the characteristic; region III: high-current characteristic*

Before advancing any further, let us first go back to the various effects of strong current densities on bipolar transistors:

– Current gain drop: this is caused by the fact that the density of minority carriers is no longer negligible compared to majority carriers of the base. The gain decreases as the inverse of the current.

– The phenomenon of the second breakdown, either thermal in origin [LET 69] (non-uniformity of the current) or electrical (modification of the electrical field by the strong densities of the minority carriers) [REY 86]. In both cases, this phenomenon of second breakdown results in focalization of the current and failure by thermal runaway.

The methodology that is normally used for modeling involves relying on existing models to which specific models for high currents and the avalanche are added. The most commonly used compact model is the simplified Gummel–Poon model [GUM 70].

Figure 4.8 presents this enriched model with a current source, i_{av}, between the collector and the base to model the avalanche of the base–collector junction for triggering. The bipolar transistor is a current source i_c that relies on two voltages: the base–emitter voltage V_{BE} and the base–collector voltage V_{BC}. The diodes model the behavior of the two juxtaposed junctions. i_f and i_r are the diffusion currents of the electrons in forward bias and reverse bias modes respectively. β_f and β_r represent the forward and reverse current gains.

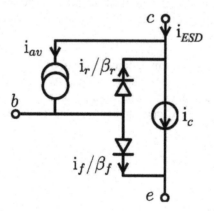

Figure 4.8. *Gummel–Poon model of the NPN bipolar transistor with a current source between the base–collector junction to model the avalanche*

The forward and reverse currents are written as:

$$i_f = i_{sf} \cdot \left[exp\left(\frac{V_{BE}}{N_f \cdot U_t} \right) - 1 \right] \tag{4.5}$$

$$i_r = i_{sr} \cdot \left[exp\left(\frac{V_{BC}}{N_r \cdot U_t} \right) - 1 \right] \tag{4.6}$$

where i_{sf} and i_{sr} correspond to the saturation currents of the diodes. N_f and N_r are the non-ideality coefficients in forward and reverse modes.

The avalanche current i_{av} of the reverse biased base–collector junction is a function of the collector current i_c and follows the relation [BER 01]:

$$i_{av} = (M - 1) \cdot i_c \tag{4.7}$$

where M corresponds to the avalanche multiplication coefficient of the carriers, given by the empirical Miller formula [MIL 57]:

$$M = \frac{1}{1 - \left(\frac{V_{CB}}{BV_{CB}} \right)^m} \tag{4.8}$$

with V_{CB} being the voltage at the terminals of the collector–base junction, BV_{CB} the breakdown voltage of this junction and $2<n<6$ an arbitrary factor dependent on the characteristics of the junction.

The disadvantage of the formulation of equation [4.8] is that it leads to serious convergence problems when the value of voltage V_{CB} gets close to that of voltage BV_{CB}. In order to circumvent this, one solution involves making an approximation of the equation by its limited development coming from the ionization integral I_n [TRE 04a]:

$$M \approx \sum_{j=0}^{N_b} (I_n)^j \tag{4.9}$$

where N_b is the number of terms and I_n is the ionization integral of the electrons, which is equal to 1 under breakdown conditions. This limited development corresponds to the sum of a geometric series of ratio I_n, leading to:

$$M \approx \frac{1 - I_n^{N_b+1}}{1 - I_n} \tag{4.10}$$

A higher number of terms, with N_b=100 as a good compromise, results in an acceptable accuracy level[2].

For the calculation of the multiplication coefficient M, the use of the theoretical expression of the ionization integral for the planar junction is justified [CHA 90] as, although the breakdown starts at the cylindrical junction, the base current of the transistor is then provided by the avalanche injection mechanism that takes place on the collector–base planar junction.

An analytical expression of the ionization integral of the electrons has been proposed by Gharbi [GHA 85] and can be used:

$$I_n = \frac{A_p}{A_p - A_n} \left[exp \left(2(A_p - A_n) \left(\frac{qN_{eff}}{\varepsilon_{Si}}\right)^3 V_{CB}^4 \right) - 1 \right]$$ [4.11]

where $N_{eff} = \frac{N_a N_d}{N_a + N_d}$ is the effective carrier concentration of the junction (N_a and N_d being the respective dopings on the P and N sides of the junction), V_{CB} the voltage at the terminals of the junction and ε_{Si} the dielectric permittivity of the silicon.

As mentioned at the start of the section, in a high-injection regime the density of the minority carriers cannot be neglected and has a strong impact on the effective carrier concentration. In order to take the high-current effects into account, an empirical approach proposed by Dutton [DUT 75] involves using an adjustment coefficient in expression [4.7] of the avalanche current, which is then defined from these measurement data. A more physical approach involves considering the impact of the minority carriers on the expression of N_{eff} [TRE 04a].

Another parameter that is strongly influenced by the effects of high injection involves the base resistance, whose value is greatly modulated by the minority carriers. This conductivity modulation is taken into account in the Gummel–Poon model. Since this resistance defines the triggering of the bipolar under the effect of the avalanche current, it must be taken out of the model of the transistor. In order to account for this triggering by avalanche current, in a vertical bipolar transistor this resistance can be broken down into an intrinsic resistance (R_{BASE} in Figure 4.9), which is located under the emitter, and an extrinsic resistance (R_{bc} in Figure 4.9), which is located in the path of the avalanche current.

---

2 It is important to note that this expression is not defined for I_n=1. However, the expression is continuous around this point and a value can be calculated. The probability that the numerical calculation precisely coincides with the singular point is infinitely low.

In order to simulate the ESD transients, the model must take into account the various non-linear capacitances of the junction (collector–base, collector–substrate and base–emitter) that are present in the intrinsic Gummel–Poon model.

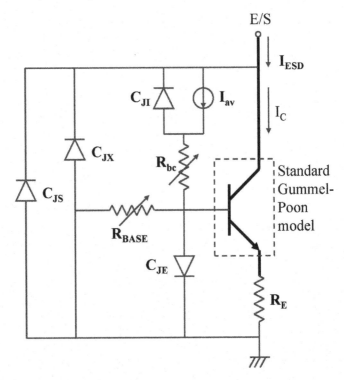

Figure 4.9. *Example of a SPICE model of a self-biased vertical NPN bipolar transistor [BER 01]. Here the base resistance is modeled by two resistances that are variable in function of the current, R_{BASE} and R_{bc} corresponding to the intrinsic and extrinsic resistances of the base, respectively. C_{JI} and C_{JX} are the non-linear capacitances of the collector–base junction, C_{JE} is that of the emitter–base junction and C_{JS} is that of the collector–substrate junction*

It is vital that these capacitances be implemented with the elements of the model that account for the operation of the structure under high-current injection. As for the base resistance, all of the parameters related to these capacitances must therefore be removed from the intrinsic model of the transistor and be moved to the elements of the macromodel. Figure 4.9 presents an example for the modeling of the bipolar transistor. These non-linear capacitances are modeled here using diodes.

4.2.2.2. Dynamic behavior

A key point for checking the effectiveness of an ESD protection is to properly model the dynamic behavior of the structure as the protection must be triggered quickly enough to protect the integrated circuit.

In the first moments of an ESD discharge, the current flowing through the transistor is purely capacitive in nature. It is mainly associated with the charge of the capacitance of the collector–base junction of the bipolar transistor. The bipolar transistor is triggered when its emitter–base voltage V_{BE} becomes greater than its threshold voltage, which is in the order of 0.6 V.

This condition is met as soon as the value of the current flowing through the base resistor is great enough. This current can have two different origins:

– the avalanche current of the collector–base junction (reverse-biased);

– the capacitive current of the collector–base junction.

The avalanche current is predominant during slow phenomena, while the capacitive current becomes more important for variations that are very rapid. The capacitive currents can therefore cause the triggering of the transistor before avalanche breakdown of the collector–base junction.

Even if the value of the biasing voltage of the emitter–base junction is great enough, the triggering of the bipolar transistor is not instantaneous. Its response time is linked to the transit time τ_B of the carriers in the base. For a transistor NPN, this is provided by the relation:

$$\tau_B = \frac{W_B^2}{m \frac{k_B T}{q} \mu_n} \qquad [4.12]$$

where μ_n is the mobility of the electrons in the base and m is a parameter that depends on the injection level, with $m = 2$ for low injections and $m = 4$ for high injections. The transit time in the base is a very important parameter for SPICE-type modeling of the bipolar transistor. Its triggering occurs at intermediate current regimes and as such its value can be chosen between the values of low and high injection and calculated using equation [4.12]. For example, for an NPN transistor with a base width of 2.3 µm, a peak doping base of 3×10^{17} cm^{-3} and doped epitaxy at 10^{16} cm^{-3}, the mean transit time is close to 1 ns.

4.2.2.3. *Extraction of the parameters of the model*

The only parameters that must be extracted to model the self-biased NPN transistor are those that characterize its operation in high-current regimes. They are mainly extracted from static measurements (DC measurements) and quasi-static measurements (TLP measurements), as illustrated in Figure 4.10.

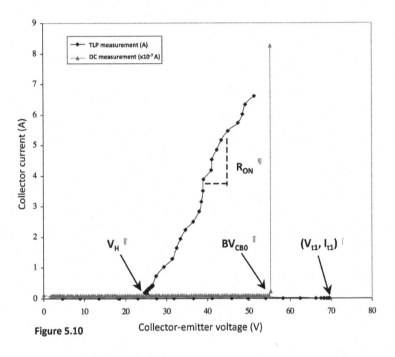

Figure 5.10   Collector-emitter voltage (V)

Figure 4.10. *Static and TLP characteristics of a self-biased NPN vertical bipolar transistor: extraction of the SPICE parameters of the ESD macromodel*

The DC measurement gives access to the breakdown voltage BV_{CB0} of the collector–base junction.

The other parameters needed for the modeling are taken from quasi-static TLP measurements:

– V_{t1}, I_{t1}: trigger current and voltage of the transistor;

– V_H: snapback voltage of the transistor;

– R_{ON}: on-resistance of the transistor;

– R_{bc}: resistance of the extrinsic base.

The same approach for modeling can be applied for a PNP bipolar transistor by using the formulation of the multiplication coefficient of the holes instead of the formulation used for electrons.

Figure 4.11 compares the results of the simulation of the modeled bipolar transistor to TLP measurements. We can see that there is a good correlation between the measurements and simulations, especially with regard to the triggering of the bipolar transistor (zoomed-in view).

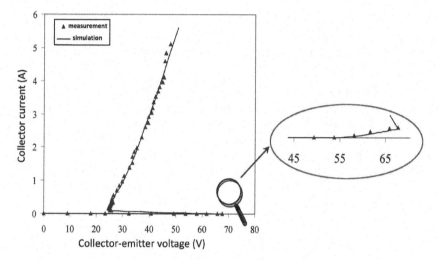

Figure 4.11. *Comparison between TLP measurement and simulation of the NPN vertical bipolar transistor on the base of the model from Figure 4.9*

4.2.3. MOS transistor model

When an MOS transistor is used as an ESD protection structure, except in the case of the large MOS transistor used as a central protection in advanced CMOS technologies, it is its parasitic bipolar transistor that is triggered to absorb the discharge current. We will therefore go back to the same modeling methodology for the bipolar transistor described in the previous section. This model of the NPN bipolar transistor is placed in parallel to the MOS model, which is taken from the technological library.

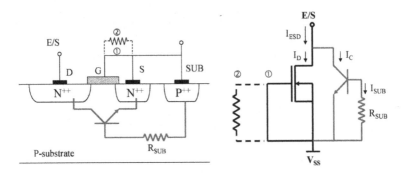

Figure 4.12. *MOS transistor in GGNMOS mode (①) or GCNMOS mode (②) with its parasitic bipolar transistor: schematic cross-section and electrical model*

This MOS transistor can be configured either in GGNMOS mode (case ① of Figure 4.12), meaning that the gate is connected to the ground, or in GCNMOS mode (case ②), meaning that there is a resistor in the gate that allows the transient ESDs on the gate to be coupled and the MOS transistor to be activated. This configuration allows triggering of the protection at a lower level of voltage that can be adjusted using the value of the resistance, as shown in Figure 4.13. In curves ① and ②, it is important to note that the snapback of the characteristic is controlled both by the bipolar effect and by the substrate effect in the MOS.

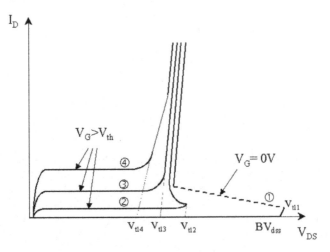

Figure 4.13. *Electrical characteristics of a NMOS transistor in the high-current regime for different gate biases*

In the case of a negative discharge, the modeling is simple as it is the diode-substrate diode that is forward-biased. However, in the case of a positive discharge, the modeling must take into account the coupling between the bipolar effect and the MOS effect.

In Figure 4.13, curve ① corresponds to the off-state mode of the NMOS transistor (gate voltage V_G being lower than the threshold voltage V_{th}). The I-V characteristic increases linearly beyond the breakdown voltage BV_{DSS} up to point V_{t11} of the snapback: this is the *first breakdown* of the component, which corresponds to the avalanching of the drain–substrate junction of the MOS transistor. This avalanche breakdown takes place at the cylindrical junction, where the radius of curvature is minimal, and the voltage BV_{DSS} is then fixed by the respective dopings of N^+ and P zones of this junction. The physical bi-dimensional simulation shows that the phenomenon of avalanche breakdown is effectively initialized at the cylindrical drain–substrate junction and also takes place mainly on the side of the gate, as the vertical electrical field contributes toward increasing the total electrical field. As a result, the breakdown appears for a voltage level that is slightly lower than that of the N^+-Substrate P diode alone. This localization of the first breakdown has been confirmed by emission microscopy experiments, which have highlighted three-dimensional effects (Figure 4.14) linked to the structure of the component. In the case where a bias voltage is applied to the gate ($V_G > V_{th}$) (Figure 4.13, curves ② ③ ④), the current of the MOS transistor contributes to the avalanche phenomenon at the drain–substrate junction and the first breakdown then appears for lower levels of the voltage (V_{t12}, V_{t13}, etc.).

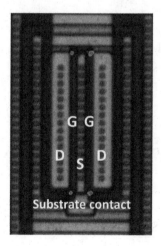

Figure 4.14. *Emission microscopy image of an interdigited NMOS transistor with two gate fingers under reverse bias in the static state (0.1 mA). We can see four corners on the side of the gate of the drains, an emission showing the initialization of the breakdown at these cylindrical junctions. G refers to the two gates, D the two drains and S the source. For a color version of this figure, see www.iste.co.uk/bafleur/ esd.zip*

Figure 4.15 presents an example of a compact modeling of an MOS transistor used as a GGNMOS or GCNMOS type ESD protection. As for the bipolar transistor, this modeling relies on the intrinsic model of the MOS, available in the current library, to which the elements relating to the high-current regime are added:

– a bipolar transistor Q_1 that models the parasitic NPN transistor inherent to the structure of the NMOS transistor. The collector of Q_1 is connected to the drain of M_1, its emitter to the source and its base to the substrate;

– a current source I_{av} that models the avalanche breakdown of the drain–substrate junction (case of a positive discharge);

– substrate resistance R_{SUB} of the NMOS transistor, which allows for the bipolar transistor Q_1 to be triggered (case of a positive discharge);

– a resistance R_D that allows the ballast resistance of the drain, as well as the resistance of its contacts, to be taken into account;

– a diode D_{SUB} that models the drain–substrate diode intrinsic to the component and lets the model operate in the forward bias mode (case of a negative discharge).

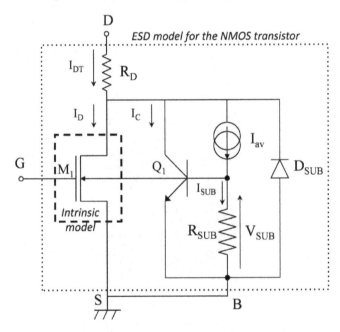

Figure 4.15. *Electrical model of an MOS transistor used in an ESD protection structure*

The expression of the avalanche current I_{av} that contributes to the triggering of the structure is similar to the one that we saw for the bipolar transistor. We only need to add the contribution of the current of the MOS transistor, which results in:

$$I_{av} = (M-1) \cdot (I_C + I_D)$$ [4.13]

where I_C is the value of the current of the base–collector diode (drain–substrate) before multiplication of the carriers and I_D is the current of the drain of the MOS transistor.

The required parameters for the compact modeling of this NPN parasitic bipolar transistor are the following:

– I_S is the saturation current;

– β_{max} is the maximal current gain;

– I_{SE} is the saturation current of the base–emitter junction;

– n_{EL} is the non-ideality coefficient of the base–emitter junction;

– I_L is the value of the collector current from which the current gain reaches its plateau value;

– I_{KF} is the value of the collector current from which the current gain drops in the high injection regime;

– R_{SUB} is the resistance determining its triggering, calculated from the measurement of the trigger current I_{t1} of the structure:

$$R_{SUB} = \frac{V_{Bint}}{I_{t1}}$$ [4.14]

where V_{Bint} corresponds to the value of the bias of the substrate at which the current gain of the parasitic bipolar transistor reaches its plateau value (~0.6 V).

The diode D_{SUB} shown in Figure 4.15 is used in the case of negative discharge in the drain in relation to the source. It is this diode that thus provides the discharge path. It must be added to the intrinsic model of the NMOS transistor as it is only represented there by the drain–substrate capacitance to which it is associated. Only two electrical parameters need to be extracted for its model. One is its saturation currents, I_S, whose value is adjusted by comparison of the results of the simulation with the experimental results. The other is the value of the series resistance R_S obtained thanks to the TLP characterization of the component in this biasing mode.

Figure 4.16 shows an example of a simulation carried out with the model of Figure 4.15. This simulation shows the impact of the gate coupling on the trigger voltage, which ranges from 13 V in the GGNMOS configuration (V_G=0) to 7 V. In a study case in Chapter 5, this modeling is used to carry out an ESD failure analysis of an IC protection strategy and to define corrective measures.

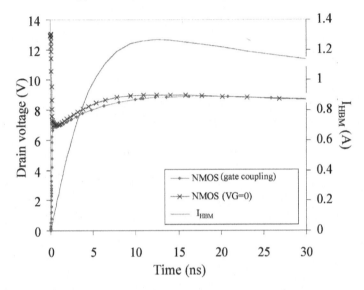

Figure 4.16. *Simulation of an HBM-type discharge of 2 kV on an MOS transistor with and without gate coupling. For a color version of this figure, see www.iste.co.uk/bafleur/esd.zip*

4.2.4. Modeling of the thyristor

The thyristor is an increasingly valued ESD protection given its excellent performance in terms of robustness. It has been kept aside for a long time due to the low snapback voltage of its characteristic (~1.2 V), which stopped it from conforming to the ESD design window, due to this low snapback voltage. Thanks to the power supply voltage of new deep submicron technologies, equal to or lower than 1 V, this limitation is no longer an obstacle to its use.

An approach for the compact modeling of the thyristor is presented in Figure 4.17. The model is broken down into three parts: modeling of the NPN transistor, modeling of the PNP transistor, and joining of the transistors to model the complete thyristor. Direct conduction involves the diode of the substrate that is modeled by a diode connected in parallel. This model is based on the elementary models of the components that we have described previously.

In this model, the resistances R_N and R_P correspond to the respective base resistances of the PNP and of the NPN. R_{ON} is the on-resistance of the thyristor. The very low values of the capacitances C are needed to obtain the convergence of the model.

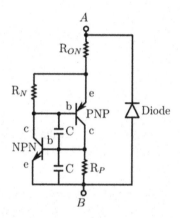

Figure 4.17. *Compact model of a thyristor*

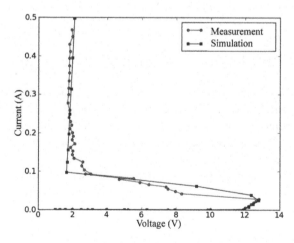

Figure 4.18. *Comparison of the TLP characteristic measured and simulated with the thyristor model from Figure 4.17 [TRE 10]. For a color version of this figure, see www.iste.co.uk/bafleur/esd.zip*

The model of the thyristor is adjusted using the following parameters:

– *on-resistance* is controlled directly by the measured value of R_{ON};

– *trigger voltage* V_{t1} is adjusted by changing the avalanche breakdown voltage of the collector–base junction of the reverse-biased NPN. The value of the voltage can be changed by acting on the avalanche multiplication coefficient M in equation [4.8] using the parameter BV_{CB};

– *trigger current* I_{t1} is adjusted via resistances R_N and R_P, thanks to the following resistances:

$$R_N = \frac{V_{BE\_PNP}}{I_{t1}}$$ [4.15]

$$R_P = \frac{V_{BE\_NPN}}{I_{t1}}$$ [4.16]

where V_{BE_PNP} and V_{BE_NPN} correspond to the voltage between the terminals of the emitter–base junctions of the PNP and NPN transistors respectively and I_{t1} is the trigger current of the thyristor;

– *holding voltage* V_H is obtained by changing the current gains β_f and β_r of the transistor as well as the non-ideality factors N_f and N_r of the junctions.

Figure 4.18 shows an example of a simulation carried out using this model, which displays a good correlation between measurement and simulation.

4.3. Behavioral simulation for prediction at the system level

Interactions within the component and between the component and the elements of the system can change the discharge paths and consequently induce unplanned failures. This chapter describes the behavioral modeling methodology that we have set up, on the one hand to deal with the issue of propagation within a system from an ESD generator up to the internal phenomena of circuits, and on the other hand to analyze the functional behavior so as to carry out susceptibility analyses of the system with regard to ESD discharge. The goal is to develop a generic methodology from the information provided by the component manufacturers, or from the measurements carried out, that can be used directly by the system designers. In this chapter, we describe how each element that makes up the system is modeled: the ESD protections, the integrated circuit, the tracks of the printed circuit board, the passive components and the discharge generators.

The systemic approach must be used as it takes into account complex interactions with a high level of abstraction. Each part of the system is considered to be independent but must include the interactions. Moreover, when using this approach it is important to simplify each of the blocks as much as possible, so as to only keep the main phenomena involved. Only following this condition is it possible to envisage complex simulations for predicting the impact of ESD on a complete system. The trade-off remains the simulation accuracy as compared to the measurement but considering the uncertainty caused by the generators for example (see Chapter 1), we estimate that 20% of error is enough to validate the approach. Some might say that 20% error for a simulation is not acceptable but to this day no other simulation is possible.

The principle behind systemic modeling is illustrated in Figure 4.19. It involves assembling elementary blocks, available in the library, like LEGO® bricks following the system topology. The component is modeled using a "black box" approach, simple enough to allow for simulations over a reduced time period but with enough information to predict a material or functional failure.

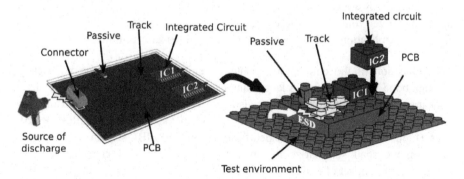

Figure 4.19. *Principle behind the modeling of a system: view of the classical system (a) and assembly into elementary blocks following the system topology (b). For a color version of this figure, see www.iste.co.uk/bafleur/esd.zip*

In our work, each elementary block was modeled in VHDL-AMS (Very-High-speed integrated circuit Hardware Description Language – Analog and Mixed Signals). VHDL-AMS is a mixed language (digital and analog), standardized in IEEE1076.1 [IEE 08]. The main advantage of this behavioral language, called HDL (Hardware Description Language), is that, just like Verilog-AMS, it allows systems to be described with multi-abstraction and multi-disciplinary models [HAM 05]. This flexibility is needed for the hierarchical assembly required in the procedure. With this approach, only the elaboration of the component models will be presented.

Before going further into the presentation of an integrated circuit model, we must first identify all of the elements that need to be considered in the model with regard to the targeted objects, meaning the prediction of material and functional failures. As defined previously, ESD can cause melting of the silicon or oxide breakdowns. When the system is destroyed by stress this is called a material failure or "hard-failure".

When the malfunction caused by the ESD is temporary and the system is able to recover function alone (without external intervention), this is referred to as a software failure or "soft-failure".

If we want to simulate the behavior of the circuit in the system, its function must be considered. From there, it is possible to consider the impact of an ESD on the generation of dysfunctions, such as clock signal losses and "RESET" for digital systems or over-voltages and jitters for analog circuits or those dedicated to radio-frequency applications. CEM specialists refer to this as susceptibility.

Connection of the chip to the printed circuit board is done via its package through bond wires and pins. A large number of studies have looked at the impact of packaging on the signal transmission, also known as Signal Integrity (SI). Currently, information on parasitic elements can be found in the standardized files, or IBIS files (Input Buffer Information Specification) [IBI 95, IBI 13]. In the following sections, we shall summarize the available information that can be useful for simulating the impact of ESD stress, especially the information provided in the IBIS files.

4.3.1. *IBIS models: advantages and limitations*

Before the middle of the 1990s, the availability of circuit description models for carrying out signal integrity simulations was very limited, with manufacturers unwilling to provide details regarding the internal structures of their inputs and outputs (I/O). The IBIS standard, also known as ANSI/EIA-656 [IBI 95] was developed to provide system designers information on the behavior of integrated circuits. Initially, the file provided no information on the parasitic elements R, L and C of the package. Since then the standard has evolved, with information on the output buffers, the protection diodes and the input and output capacitances. The IBIS files allow various effects to be predicted, such as over-voltages, diaphony, delays and impedance losses that could degrade the electrical signal [IBI 95, CUN 96, HAL 09, HYA 02].

The IBIS file contains information on each of the different pins of the integrated circuit. The main pieces of information provided in this file are:

– the values of elements R_{pkg}, C_{pkg} and L_{pkg} corresponding to the package and the wires connecting the chip to the package;

– the value of the capacitance C_{comp}, which is the equivalent capacitance of the input or output level tied to the transistor, to the metallic connection tracks and to the capacitance of the pad;

– the I(V) tables of the "power clamp" and "ground clamp" ESD protections corresponding to the protections located between the pad and the power supply rails V_{DD} and V_{SS} respectively. It must be noted that the "power clamp", at least from the IBIS point of view, is not the protection between V_{DD} and V_{SS} as we know it in the ESD community;

– the I(V) tables called "Pull Up" and "Pull Down", corresponding to the output transistors. In the case of a circuit in a CMOS technology for example, these curves correspond to the PMOS and the NMOS of the output stage. The I(V) characteristics are obtained for $|V_{gs}| = V_{DD}$;

– the four V(t) tables that provide information on the output dynamics. These waveforms correspond to the rising and falling transitions of the Pull Up and Pull Down tables of the output stage.

The equivalent electrical diagrams of an input and an output are illustrated in Figure 4.20(a) and (b). The ESD protections are represented by diodes.

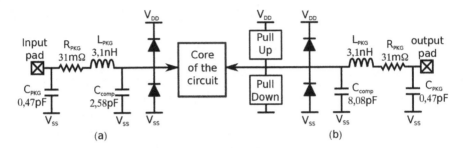

Figure 4.20. *Electrical diagrams of the IBIS models of an input (a) and an output (b) – extracted for a 74xx00 component.*

According to the description that we have just made, the IBIS files appear to contain all of the required information, including information on ESD protections. The only one missing is the protection between V_{DD} and V_{SS}. Studies have been carried out by N. Lacrampe and N. Monnereau [LAC 08, MON 11a] to see whether IBIS models could be used to predict ESD disturbances taking place at the I/O pins of an integrated circuit.

4.3.1.1. *Importance of the package elements and of the internal passive elements*

To illustrate the role of IBIS files, let us take the example of the IBIS file for passive elements of the circuit 74LVC04A. This file contains the models of the four types of package: SOP, SSOP, TSSOP and SOIC. The extracted values of the elements R_{pkg}, L_{pkg}, C_{pkg} as well as the capacitance C_{comp} are given in Table 4.2. We carried out a frequency analysis of the four packages in order to determine their cut-off frequencies. The capacitance C_{comp} depends on the chip alone and therefore does not vary as a function of the package.

When the ESD protections have not been triggered, they are considered to be open circuits and have no influence. The package is directly connected to the transistors of the input stage. The impedance of this stage is very big, in the order of several megaohms, and as such it can be considered to be an open circuit. For the simulation, we introduced a high value of impedance in the order of 10 MΩ. The cut-off frequency for −3 dB is called F (Zhigh) in Table 4.2.

	F (Zhigh)	F (Zlow)	R_{pkg}	L_{pkg}	C_{pkg}	C_{comp}
SOP	2.49 GHz	35 MHz	45 mΩ	3.796 nH	0.53 pF	2.58 pF
SSOP	2.58 GHz	42 MHz	44 mΩ	3.55 nH	0.42 pF	2.58 pF
SOIC	2.76 GHZ	49.3 MHz	31 mΩ	3.109 nH	0.47 pF	2.58 pF
TSSOP	3.2 GHz	67.5 MHz	32 mΩ	2.27 8 nH	0.31 pF	2.58 pF

Table 4.2. *Cut-off frequency obtained for different types of package*

When the protection is triggered, the value of its equivalent resistance becomes very low. For the simulation of the bandwidth, we placed a resistance of 1 Ω at the output. The cut-off frequency (Zlow) then drops to less than 100 MHz for all of the packages.

The impact of the parasitic elements of the package on the current paths is considerable. In terms of the ESD stress, the low value of the resulting equivalent inductance, 3.6 nH, plays a major role during strong discharge current transitions following the law U = L.di/dt. For an IEC stress of 8 kV generating a current peak of 25 A in less than 1 ns, this inductance introduces a voltage peak of about 90 V across the terminals of the capacitance. An example of this effect has been analyzed [MON 10] and shows that the combination of an external capacitance and the package inductance results in a significant change in the discharge path of the component. The injection of a single TLP pulse leads to the appearance of current and voltage peaks. This case study is further developed in Chapter 5.

In the following section, we shall evaluate the impact of internal capacitances (C_{comp}). The electrical schematic used for this study is shown in Figure 4.21 [LAC 08]. The ESD pulse is injected onto line 1 to create a disturbance through coupling on line 2. The amplitude of the noise generated, as well as its dynamics, are measured and compared to the simulation.

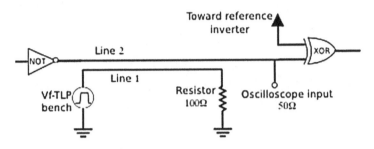

Figure 4.21. *ESD disturbances by line coupling
on the I/O output leads of the circuits*

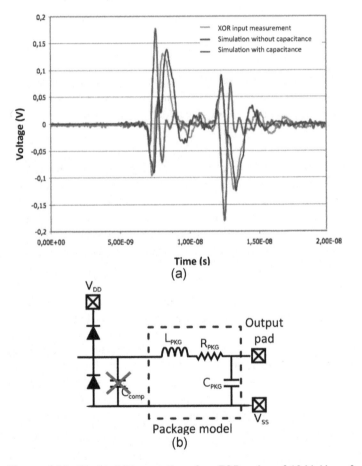

Figure 4.22. *Study of the coupling of an ESD pulse of 10 V. Use of
the IBIS model without input and output capacitances of the circuits.
For a color version of this figure, see www.iste.co.uk/bafleur/esd.zip*

In order to properly understand the waveforms visualized at the input of the XOR port, several simulations have been carried out, with or without taking the capacitance C_{comp} into account. The pulse VF-TLP has a rise time of 300 ps, a length of 5 ns and a voltage level of 10 V to not trigger the protection structures. Figure 4.22 shows the measured and simulated voltages (a) as well as the models used to simulate the input and output stages of the circuits (b). The simulation without capacitance reveals transitions that are too rapid. The internal capacitance C_{comp} plays an important role in the dynamic representation of the inputs and outputs. It is included in the following simulations.

4.3.1.2. Role of diodes delivered by IBIS

Going back to the previous case, the amplitude of the pulse is increased in order to trigger the protection diodes. When the diodes are taken out of the simulation, considerable over-voltages appear (Figure 4.23) highlighting the importance of considering the protection elements in simulations of high amplitude transient phenomena.

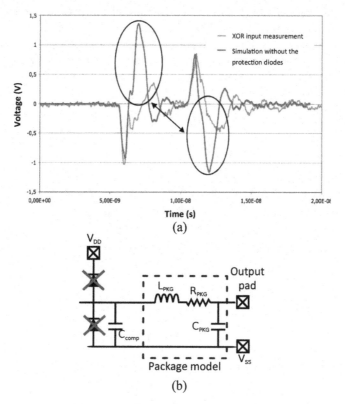

Figure 4.23. *Study of the influence of the protection diodes on the waveform of the signal (injection of an ESD pulse of 100 V). For a color version of this figure, see www.iste.co.uk/bafleur/esd.zip*

Circuit	Parameters	Power clamp diode	GND clamp diode
INV	IBIS	RS=10 Ω; BV= 10 V; N= 1.5	RS=6 Ω; BV= 10 V N= 1
	Optimized	RS=6 Ω; BV= 10 V N= 0.8; CJ0 = 2.0 pF	RS=6 Ω; BV= 10 V; N= 1; CJ0 = 1.0 pF
XOR	IBIS		RS=6 Ω; BV= 7.5 V; N= 1
	Optimized		RS=6 Ω; BV= 7.5 V; N= 1; CJ0= 1.0 pF

Table 4.3. *Values of the various parameters for the ESD protection diodes of each circuit*

In the following study (Figure 4.24), we carried out two simulations. One of them used diodes extracted from IBIS files and modeled classically using SPICE. The second simulation is carried out using diodes that we optimized (also described using SPICE), called "optimized diodes". The model is based on the TLP measurement (quasi-static measurement) of these same diodes. Table 4.3 shows the main parameters of the SPICE model.

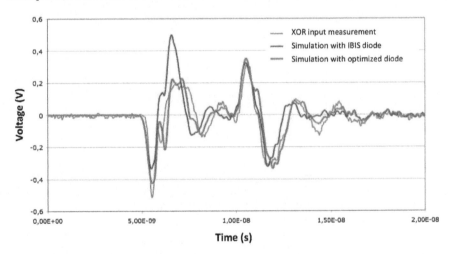

Figure 4.24. *Study of the coupling of an ESD pulse of 50 V. The IBIS model is used with parameter values for the ESD protection diodes provided by IBIS and optimized diodes from quasi-static measurements. For a color version of this figure, see www.iste.co.uk/bafleur/esd.zip*

The comparison between the measurement and the simulations for the two types of model is shown in Figure 4.24. We can see the importance of using diode models extracted from quasi-static measurements. The I(V) characteristics of the protections are given by IBIS for a range of voltages from $-1*V_{DD}$ to $2*V_{DD}$, which is too low for studying ESD events. Moreover, static and quasi-static measurements do not provide the same information, especially if the circuit is protected with dynamically or snapback-triggered structures. A quasi-static measurement (TLP) carried out between the pin V_{SS} and the input of the circuit is compared to the I(V) characteristic provided by IBIS in Figure 4.25. The value of the on-resistance of the IBIS diode is far too big in comparison to the real behavior of protection structures during discharge. This is because the IBIS model does not characterize internal ESD protection structures and only a static behavior is dedicated to signal integrity analysis.

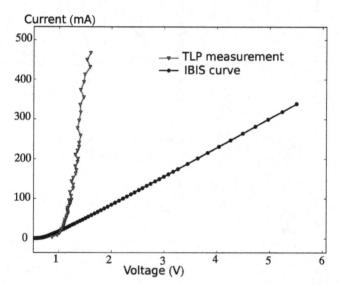

Figure 4.25. *Comparison of a TLP measurement and of an IBIS I(V)*
characteristic of the structure located between the lead VSS
and the input of the circuit

4.3.1.3. *Role of the output buffers provided by IBIS*

A study carried out by N. Monnereau shows that pull-up and pull-down stages must be considered, as they are involved in the conduction of the ESD current [MON 10]. During a stress on the output, as shown in Figure 4.26, a discharge path is created through the pull-down. The amount of deviated current corresponds to the amount specified in the IBIS files. This case study is shown in detail in section 5.3.1 of Chapter 5.

A study carried out by S. Giraldo highlights the destruction of an integrated circuit used for sound amplification in mobile phones through this phenomenon [GIR 10]. In this study, the component is placed in its final application and the stress is injected onto the audio output of the operational amplifier. The high value decoupling capacitances C_{dec} and C_{out} (1 µF) allow proper operation of a charge pump (LDOP and LDON) (Figure 4.27). In this final configuration the protection structures, which are active protections, are not triggered as the external capacitances maintain fixed potential at their terminals. Analyses have been carried out to determine the robustness level depending on the power supply. Table 4.4 shows that robustness decreases drastically as the power supply increases.

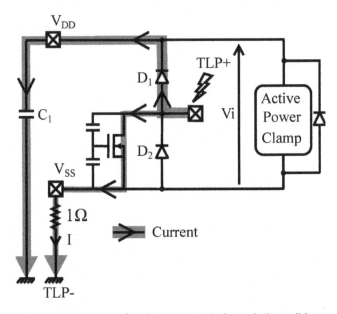

Figure 4.26. *Appearance of a discharge path through the pulldown stage*

When the system is being powered, the NMOS transistors are in the on-state. Through the capacitance C_{GD}, the ESD stress induces an increase in the gate voltage of the transistor M_0, increasing its conduction beyond the safe operating area (SOA). The TLP measurement that follows (Figure 4.28) shows that when the gate of the NMOS transistor is biased to 5 V, the failure level appears when V_{DS} goes beyond 5 V. The simulation, however, gave a voltage excursion of 7.5 V for this same level of bias.

Power supply voltage	0 V	3.8 V	5.2 V
Failure level	7 kV	2 kV	1 kV

Table 4.4. *Values of the different parameters for the ESD protection diodes of each circuit*

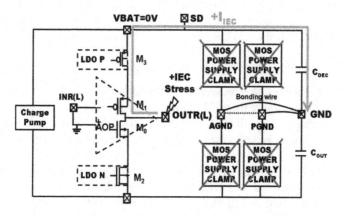

Figure 4.27. *Simulation of the current path during an IEC discharge applied to the amplifier output [GIR 10]*

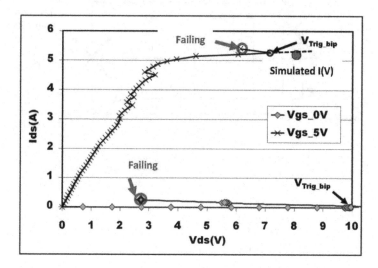

Figure 4.28. *TLP characteristic IDS (VDS) of the NMOS M0 transistor in function of the gate bias VGS = 0V and VGS = 5V [GIR 10]. For a color version of this figure, see www.iste.co.uk/bafleur/esd.zip*

To carry out such a simulation and predict the failure it is imperative to have information on the output transistor (the pulldown), the value of the coupling capacitance of the gate and measurement data quantifying the SOA (Safe Operating Area). This study was completed thanks to many complementary measurements to validate the simulated approach. Given the results, it seems important to have information on the size of the output buffers as provided by IBIS, even if the IBIS cannot directly provide this information. However, in order to predict the behavior of these buffers in response to an ESD stress, we would also need the capacitance that couples their control to the output, as well as the safe operating area (SOA).

4.3.1.4. Conclusion on IBIS files – improvements

Generally speaking, the representation of diodes given by IBIS is not enough to represent the quasi-static reality of ESD protections (dynamic resistance, snapback, dynamic triggering and amplitude). Moreover, all of the current paths taken by an ESD in the component are not taken into account. The schematic representation of a centralized protection strategy showing the missing paths not defined by IBIS is given in Figure 4.29. We can see that the power supply protection between the pins V_{DD} and V_{SS} is not defined. If a discharge takes place between an output (OUT) and V_{SS}, there is no structure or path that can evacuate the current toward V_{SS}. This path does exist however and must be taken into account.

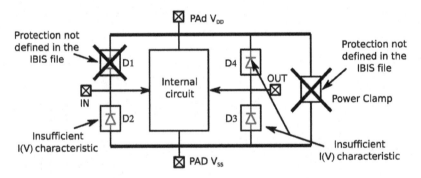

Figure 4.29. *Highlighting of the drawbacks of IBIS, for circuit 74LVC04A, in the diagram of a typical centralized protection strategy used in CMOS technology*

The integrated circuit model that we propose preserves the values of the passive elements of all the pins (R_{pkg}, L_{pkg} and C_{pkg}), as well as the equivalent capacitance (C_{comp}) given by IBIS. The functional model of the circuit reuses the I(V) and V(t) tables of the primordial PullUp and PullDown to carry out functional simulations [MON 11a, MON 11c]. Table 4.5 that follows summarizes the IBIS elements that can be preserved for ESD simulations and those that need to be replaced.

Element	Included in IBIS	Role	Improvement
Parasitic elements of the package	Yes	Inductance: – limitation of abrupt current excursions; – modification of the current paths: preservation of the instantaneous potentials; – better convergence of the simulator during high amplitude stress. Resistance and capacitance: impact on the leakage currents during transients	Keep without changes
Buffer	Yes	Allows determination of the current that can flow through them during a stress	Lacks: – coupling capacitance toward the gates; – safe operating areas (SOA) for the buffers.
Capacitance C$_{comp}$	Yes	Represents the dynamics of the inputs and outputs and the equivalent capacitance of the protections	Keep without changes
Input output diodes	Yes	Clamping during high injection stress	Change entirely: – Quasi-static measurements for their characterization. – Extension of operating zones to very high currents. – Easily interchangeable behavioral description.
ESD protection between power supplies	No	Essential for predicting the current path in the component	To be added imperatively, following the previous recommendations
ESD protection between the grounds	No	Essential for predicting the current path in the component	To be added imperatively, following the previous recommendations

Table 4.5. *Summary table of the IBIS elements to be modified or completed for system level ESD simulation*

4.3.2. Setting up a power supply network

The structure of the power supply network is needed to simulate the circulation of the current in the component. The setup must go back to the topology of the power supply network of the component, as specified in the ICEM model [IEC 10]. The ICEM model thus defines the PDN (Passive Device Network) needed for simulation of the power supply network for CEM emission. The network is made up of passive elements, hence its name.

The notion of "Power Delivery Network", using the same acronym, allows active elements such as ESD protection structures to be placed onto the power supply. In order to overcome this confusion, standardization committees are considering the unification of the notions used. The idea is to define a non-linear network called NLN, which would include all of the non-linear elements of the protections. The important step is to define an extraction standard for the ESD protections, which would feed the description of the NLN.

An approach that is already generally agreed upon is to rely on the TLP characterizations that we shall develop in the following sections.

4.3.3. Extraction of parameters from measurements

To extract information on ESD protections, static and quasi-static (TLP) measurements must be carried out between sensitive pins. Thanks to these measurements, the protection strategy can be analyzed. The comparison between static and quasi-static measurements allows identification of the ESD structures that use a dynamic triggering circuit calibrated to the rapid transitions of ESD [SEM 08, VOL 99, BER 99]. Only the quasi-static characteristic is used to model the behavior of the protections in the high injection regime.

4.3.3.1. Approach for compact modeling

A first approach for modeling would involve developing a physical model of the component, depending on the structure identified, where the parameters would be adjusted to correspond to the characteristics measured. Once the protections are identified, the compact models presented in section 4.2 can be used. This gives simulation results that are precise. On the other hand, if the component for which the model being developed has a large number of pins, things become more complex. The models of all the protections of the component must be adjusted and, importantly, the electrical simulation carried out must contain a large number of components, which can be detrimental to the simulation time during rapid transients.

In the case of the CMOS inverter presented here, three types of structure have been identified: diodes, a NMOS transistor used as a central protection with its associated RC trigger circuit and a thyristor. The protection strategy, reconstituted with the help of measurement, is illustrated in Figure 4.30(a) and (b).

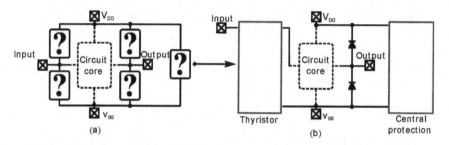

Figure 4.30. *Centralized protection strategy in CMOS technology: no information on the protections (a) and protections identified through measurement (b)*

The static and TLP measurements carried out between the input and the V_{SS} pins allowing us to identify the thyristor are given respectively in Figure 4.31(a) and (b). The static measurement is carried out with a current limit of 50 mA to avoid destruction of the circuit. The TLP measurement is carried out by adding a 500 Ω resistor in series, limiting the current and allowing us to accurately extract the triggering and then the snapback of the characteristic. The structure starts to be triggered at 12 V. When the voltage reaches V_{t1} = 12.8 V for a current of I_{t1} = 30 mA, the characteristic snaps back to a holding voltage and current of V_h = 2 V and I_h = 110 mA. The on-resistance R_{ON} is in the order of 1 Ω. The snapback at 2 V is typical for a thyristor type structure.

Static measurements must also be carried out. They allow determination of the ESD structures that use a dynamic triggering circuit calibrated on the rapid ESD transitions. The static and TLP measurements carried out between V_{DD} and V_{SS} pins help identify an MOS structure with a dynamic trigger and are shown respectively in Figure 4.32(a) and (b). These measurements were carried out under the same conditions as for the thyristor. When the structure is measured in static, the NMOS transistor is blocked. The characteristic obtained corresponds to the reverse-biased diode of drain N/substrate P. However, for the TLP measurements, the rapid rising edge of each pulse triggers the NMOS transistor. The measurement obtained corresponds to the $I_D(V_{GS})$ characteristic of the NMOS transistor. The trigger threshold is V_{t1} = 0.7 V.

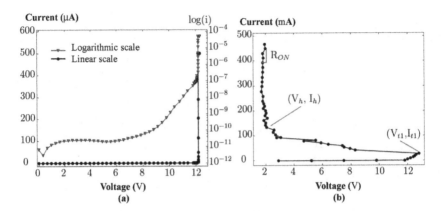

Figure 4.31. *Static measurements (a) and TLP measurements (b) between the input and VSS pins of the circuit: identification of a thyristor*

Based on these measurements, the equations from section 4.2 can be used to generate the compact models of the structures identified. The models developed can be used to determine the errors introduced by another modeling approach, or behavioral approach, where few physical notions are needed to represent the protections. These two techniques have been analyzed and compared [MON 11a] but we shall focus here on the behavioral description while presenting the improvements in performance, precision and calculation times compared to the physical models.

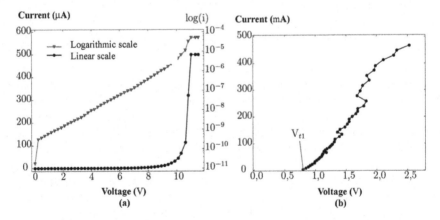

Figure 4.32. *Static measurements (a) and TLP measurements (b) between the VDD and VSS pins of the circuit: identification of an NMOS with dynamic RC trigger circuit*

4.3.3.2. *Behavioral modeling approach*

This technique presents the disadvantage of no longer being able to associate a physical reality to the protection. However, the high level of abstraction introduced allows direct elaboration of the model into the language of behavioral description. This approach, called "systemic", also presents the advantage of guaranteeing the intellectual property of integrated circuits. Finally, physical abstraction lets us describe highly complex functions that can be simulated very rapidly.

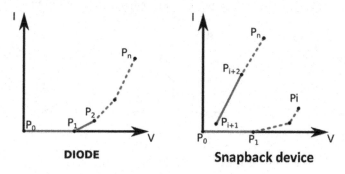

Figure 4.33. *Generation of a library of behavioral models of protections using a piecewise linear description*

The approximation of the quasi-static I(V) characteristic is done using line segments (Figure 4.33). The choice of this technique can be justified by the simplicity of its application. On the other hand, it does generate discontinuities over several values, which must be dealt with in order to ensure convergence of the simulation.

The I(V) characteristics of ESD protections usually present a limited number of identifiable aspects. Concretely, the characteristic is either a diode characteristic or it contains a snapback. This statement has led to the creation of a small library of behavioral models that covers a large number of cases. Two I(V) curve approximations have been implemented, for which the coordinates of the inflexion points P_i (Figure 4.33) can be parameterized by the user. For all of the measurements that we present in our studies we use a TLP bench that generates a pulse duration of 100 ns with a rise time of 300 ps. Under these conditions, the validity of the extraction of the parameters can be debated given that the rise times and the pulse durations can have an influence on the characteristics depending on the type of protections encountered. Most articles published [MER 12, SCH 12, BER 12, ORR 13] use a similar pulse duration and yield good predictions of the robustness of the systems studied.

4.3.3.3. Methodology for behavioral modeling

In order to explain the behavioral approach, we will take the example of the modeling of a forward-biased diode, as in Figure 4.34(a). From a TLP measurement, the behavior of the diode is linearized. Only two states are used, each corresponding to an analog equation:

– state 0: I=0;

– state 1: I=U.R_{on}, where R_{on} represents the dynamic resistance of the protection.

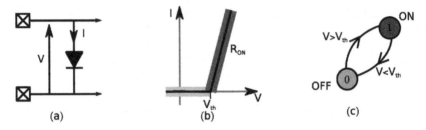

(a)   (b)   (c)

Figure 4.34. Schematic representation (a), ideal I(V) characteristic of a forward-biased diode (b) and state diagram of the behavior of the diode (c)

This type of model can easily be implemented in any electrical simulator. However, the abrupt switching from one state to the other can easily introduce strong discontinuities in the signal or its derivative, resulting in considerable problems in terms of simulator convergence. This is particularly true in the case of ESD simulations, where the current transitions are very rapid (a few 10 A/ns). The value of the equivalent capacitance C_{comp}, given by IBIS and connected in parallel, helps avoid voltage discontinuities between the terminals of the protection. It is needed in order to obtain the convergence during state changes.

4.3.3.4. Application to a snapback structure

To illustrate the outline of the modeling process in this precise case, we shall present a TLP measurement carried out in a thyristor, as shown in Figure 4.35. The inflexion points of its I(V) characteristic are extracted and used as parameters. Six parameters, made up of a current/voltage (V_X, I_X) couple, are needed to implement the complete model of the thyristor. These are reported in Table 4.6 and in the TLP curve in Figure 4.35. Four modes, highlighted in the TLP measurement, are identified: conduction step (0), reverse conduction before the snapback (1), snapback (2), forward conduction (3). There is no intermediate state between state (1) and state (2). The points observed between these two states in Figure 4.35 are measurement artifacts.

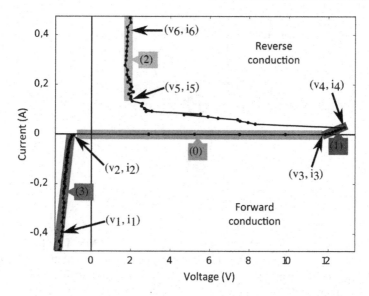

Figure 4.35. *TLP measurement between the input and VSS pins and characteristic inflexion points*

Parameters	1	2	3	4	5	6
Voltage (V)	−1.5	−0.7	11.8	12.9	1.9	2.2
Current (A)	−0.4	0	0	0.05	0.1	0.4

Table 4.6. *Thyristor parameters extracted from the TLP curve*

Each of the four operating modes of the thyristor corresponds to a linear equation whose directing coefficients are calculated from the parameters in Table 4.6. The state diagram established for this thyristor is shown in Figure 4.36. The snapback condition corresponds to state 2 when the voltage is greater than V4, and no return condition between state 2 and state 1 is possible. The structure can only go from state 2 to state 0 if the value of the voltage drops below V5. By adapting the state graph, we can follow the chronological operation of the protection, as it is clear that when the current drops, there is no over-voltage.

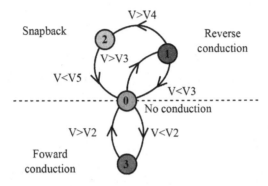

Figure 4.36. *State diagram of the thyristor*

The model of the component is translated into VHDL-AMS. This language helps describe not only digital circuits but also analog ones. The calculation environment used therefore has two simulation cores: one is purely digital, which is event-related; another is analog, which is temporal. The state diagram of the transistor is re-transcribed in the digital nucleus through a process that allows the lines of code to be executed sequentially. The flowchart corresponding to the thyristor is shown in Figure 4.37. The state of the structure is determined as the process is executed. The process then waits until one of the output conditions of the detected state becomes true (loop). At this moment, a flag marks the new state of the thyristor. The process is re-executed and continues to wait for new events. The states thus defined allow selection of the functional equations.

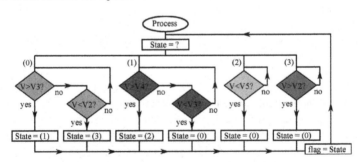

Figure 4.37. *Flowchart of the process implemented for the modeling of the thyristor*

Generally, a linear behavioral model can be extracted from a TLP measurement between two pins. Each segment is associated with a function mode or state in the digital nucleus, which allows an analog equation of the analog nucleus to be chosen in agreement with the behavioral model of the structure.

Conditions regarding the derivatives of a signal (for example, dV/dt) can also be established. For example, to model the triggering of the dynamic NMOS transistor (see section 0), an equivalent RC circuit can be used whose time constant has been adjusted. Thanks to the use of behavioral languages, this equivalent triggering circuit can be omitted. To change the state of the protection structure, just like for the triggering of a "Power Clamp", we must implement a condition over the derivative of the voltage between its terminals [MON 11a, GIR 13].

Table 4.7 summarizes the advantages and disadvantages of the two modeling approaches, compact and behavioral. The behavioral modeling approach is best adapted toward analysis of the impact of ESD at the system level.

	Compact models	Behavioral models
Advantages	– Close to the physics of the component – Parameterized model	– Simple parameter extraction – Rapid development and tuning – Good robustness with regards to convergence problems – Rapid calculation time
Disadvantages	– Physical parameters that are hard to adjust – Difficult to converge for very rapid transitions	– Models that do not allow for the optimization of the protections

Table 4.7. *Comparison of compact and behavioral modeling for analysis of the impact of ESD at the system level*

4.3.4. Application to system modeling

This paragraph looks at the effort invested into creating a purely behavioral modeling methodology. This makes perfect sense for obtaining failure predictions, both material and functional, of complex systems that integrate several integrated circuits. We have shown that the elements that make up the electronic board, as well as the environment in which the system is placed, play an important role in the propagation of the ESD signal. Special care was needed to model the integrated circuit. For the modeling of the protections the only data provided, except that included in the IBIS, are the inflexion points of the characteristics, measured externally and quasi-statically. The representation of states coded in VHDL-AMS in the form of graphs is up to the user. It was our choice to use the graphs to easily solve problems of convergence, as seen in various publications that go into details on the behavioral approach [MON 10, MON 11a, MON 12, MON 13a, BES 11, CAI 12, CAI 13, CAI 15a, BAF 13, BÈG 14a, GIR 14, ESC 16a]. Modeling by segment can also be implemented in other descriptive languages such as Verilog-

AMS [VER 14]. Such a component model is able to predict both material and functional robustness. The choice of failure criteria is vital and can be implemented at the level of the protections (Wunsch & Bell type material robustness [WUN 68]) or at the heart of the function of the component (functional robustness). The advantage of having simplified the model is to make it interchangeable, thus protecting intellectual property. For the modeling of the complete system, this component model is inserted into the electrical diagram of the electronic board that integrates the passive elements and the interconnection lines. Finally, the environment is added to take into account the interactions between the measurement devices, the generators and especially the grounds. All of the models developed are assembled in a structural and hierarchical manner, following the topology of the system being modeled.

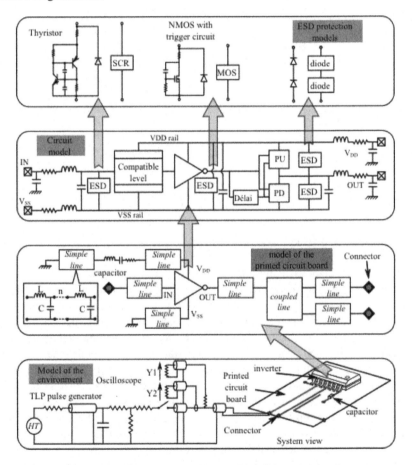

Figure 4.38. *Hierarchical assembly of the models of the library according to the topology of the system*

An illustration of the assembly of all of the elements is given in Figure 4.38. At the bottom, the first rectangle represents the test environment, made up of a TLP bench, connecting wires and current and voltage probes connected to the oscilloscope. This test environment is linked to a connector of the board under test. By going down one level in the hierarchy we can see the model of the printed circuit board, including models of the lines, models of the passive components and the integrated circuit block. The next level in the hierarchy presents the model of the integrated circuit, made up of elements extracted from the IBIS models and from the ESD protection blocks. Finally, the last level corresponds to the models of the ESD protections integrated in the circuit that are described behaviorally.

4.3.5. Validation of the behavioral models through the system approach

The systemic modeling approach developed previously has been validated in several case studies. In the previous section, these works were synthesized as part of the goal of simulation. Firstly, the quasi-static analysis is validated before looking at transient simulation. The failure levels are then tackled, with the introduction of criteria of material and functional failure in the models. Only the most marked results are presented, as well as an analysis of performance levels during simulation.

4.3.5.1. Quasi-static analysis

The validity of the behavioral models has been verified [MON 12] by analyzing all of the discharge paths that involve more than one protection. TLP simulations using both behavioral and compact models were compared with measurements in all possible discharge configurations in a combinational logic circuit. The models were elaborated for an input, an output and for the power supply pair V_{DD}, V_{SS}, which in other words represent 12 discharge paths. These were then compared to the measurements so as to verify satisfactory correlation (Figure 4.39). It must be noted that the stress configurations $V_{DD} \rightarrow$ IN, $V_{SS} \rightarrow$ OUT as well as IN $\rightarrow V_{DD}$ are not shown in the figure for the sake of clarity, as they would otherwise be superimposed over the configurations $V_{DD} \rightarrow$ OUT, $V_{SS} \rightarrow$ IN and IN $\rightarrow$ OUT. In any case, we can see there is a good level of correlation between simulation and measurement.

The relative error of the voltage introduced by the behavioral and compact models is shown in Figure 4.40. We have chosen to represent this error as a voltage and not as a current. Indeed, the current that flows through the structure once the quasi-static state has been established is fixed by the ESD discharge current. Only the voltage at the terminals of the structure can change. The error is calculated for each current level, according to the following relation:

$$error = \frac{\left| voltage_{simulated} - voltage_{measured} \right|}{voltage_{simulated}} \qquad [4.17]$$

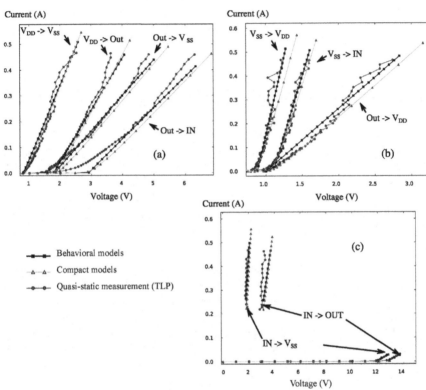

Figure 4.39. *Comparisons between compact modeling, behavioral modeling and measurement of the TLP stress for various discharge paths*

As we can see in Figure 4.40, in any case the error introduced by the models is lower than 20%. It is comparable to the value obtained with SPICE-type compact models.

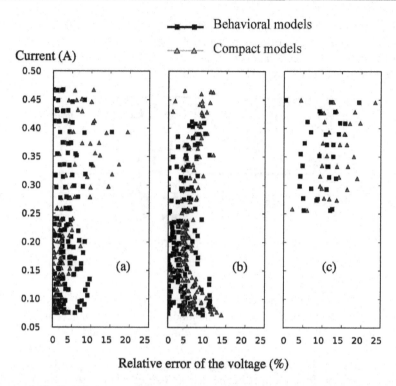

Figure 4.40. *Relative error of the voltage introduced by the behavioral and semi-empirical models for the structures: (a) diodes, (b) GCNMOS and (c) thyristor*

4.3.5.2. *Transient simulations*

The model must be validated in the transient regime for different stress systems such as those in the standards IEC61000-4-2, HMM or ISO 10605.

Different case studies have been developed with the goal of demonstrating the validity of the behavioral models for the transient simulation [BÈG 14a, ESC 16a, ESC 16b]. To put this method in place, we started to work on simple cases such as inverters, flip-flop circuits and then more complex components such as LIN transceivers dedicated to onboard automobile networks. From the moment the system is correctly modeled, it is possible to achieve a precision of less than 20%.

The example presented below is the result of a collaboration with EADS-IW and Continental, with the goal of demonstrating the validity of behavioral models in the transient regime. The electrical diagram of the regulator board developed during this project is presented in Figure 4.41.

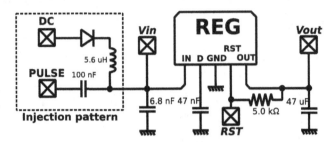

Figure 4.41. *Electrical diagram of the test board for the automobile regulator [BÈG 14b]*

After extracting the protections between IN, OUT and GND by TLP measurements, we developed the complete model of the component (Figure 4.42). In parallel to the protections between these three pins, we added their respective capacitances. This is needed in order to carry out a temporal simulation and to obtain a linear evolution of the voltages at the terminals of the protections. As mentioned previously, the effect of the discontinuities of the behavioral models on the simulation is smoothed out by the addition of capacitances. The electrical diagram reveals two internal blocks that were implemented in order to carry out a functional simulation of the output-regulated voltage and of the generation of a RESET signal by the component [BÈG 14a].

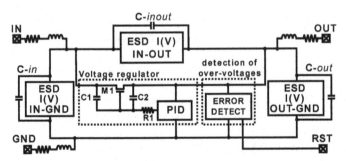

Figure 4.42. *Complete model of the automobile regulator used for the simulation [BÈG 14b]*

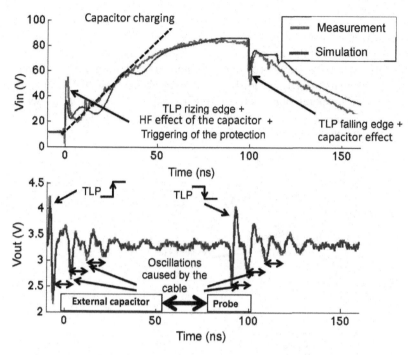

Figure 4.43. *Comparison of the measurement and simulation of the dynamic input and output voltages during a TLP stress of 600 V onto the input in the powered configuration [BÈG 14a]. For a color version of this figure, see www.iste.co.uk/bafleur/esd.zip*

The next step involves validating the dynamic simulation with a TLP injection (Figure 4.43). This step is crucial, as beyond simply validating the model, it also ensures that all the elements of the system (cable, PCB, high frequency parameters of the passive elements) are well-modeled. As we have seen previously, each element influences the measured voltage and current waveforms. In the measurements and simulations presented in Figure 4.43, the system is supplied by 12 V and the regulator provides a voltage of 3.3 V at the output. The input onto which the injection is made and the regulated output are compared at the same time. The voltage at the input increases greatly and the network of protections of the component carries out its role by limiting the charge of the external capacitance. The output oscillations are strongly linked to the elements of the measurement bench. The first peak is linked to the capacitive coupling between the input and the output. The frequency of the oscillation observed is caused by the length of cables used. This experimental phenomenon is well represented by the simulation.

From the validation of the system simulation it is then possible to consider a simulation with other types of stress, such as the IEC61000-4-2 presented in Figure 4.44. The prediction of the signal waveforms (input and output) is relatively good in terms of the uncertainty introduced into the component, and especially with regards to the environment, which is a lot harder to control than that of the TLP bench. However, such temporal simulations are precise enough to carry out a robustness analysis. The simulation also reliably reproduces the behavior of the voltage regulation function during stress (voltage V_{out}).

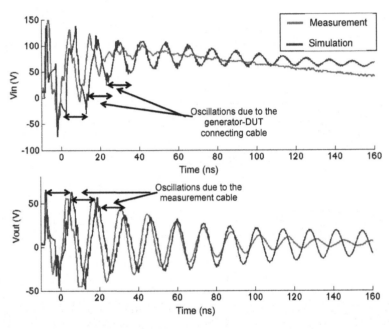

Figure 4.44. *Comparison between the measurement and simulation of dynamic input and output voltages during an IEC 61000-4-2 stress of 8 KV [BÈG 14a]. For a color version of this figure, see www.iste.co.uk/bafleur/esd.zip*

We have shown that empirical and behavioral models provide more or less the same results during a quasi-static and transient simulation. The models were also validated [BÈG 14a] for other types of transient stress such as the HMM and the stress from the standard ISO 7637. The performance of these models in terms of calculation time is also an important factor for choosing the model to use during the simulation.

4.3.5.3. *Performance in term of calculation time*

To evaluate this performance level, a comparative study was carried out by system simulation with behavioral models and compact models. The internal protection between V_{DD} and V_{SS} of the simulated component is successively an SCR structure and a Power Clamp. An external decoupling capacitance (25 nF) is placed between the V_{DD} and V_{SS} pins of the circuit. The injection is done on the output of the component. The diagram of the simulation principle is shown in Figure 4.45.

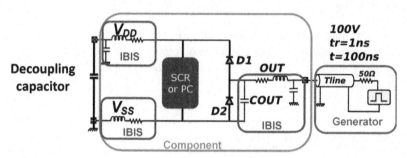

Figure 4.45. *Electrical diagram used to compare the performance of the behavioral and compact models. For a color version of this figure, see www.iste.co.uk/bafleur/esd.zip*

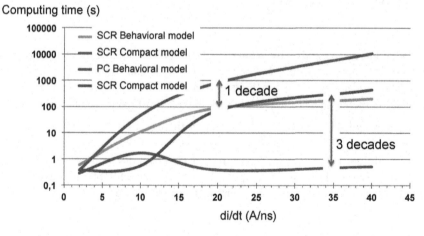

Figure 4.46. *Simulation results comparing the performance of behavioral and compact models. For a color version of this figure, see www.iste.co.uk/bafleur/esd.zip*

The base structures, thyristors and active NMOS onto which we sent the TLP pulse, of current transitions of 2–40 A/ns, are analyzed. The impact of the $\frac{di}{dt}$ on the calculation time for the two types of modeling is shown in Figure 4.46. When the transition is relatively slow, there is no real difference. However, as soon as the transitions are rapid, the compact model starts to take up a lot of calculation time, with a ratio that is far greater than 10 beyond 20 A/ns for the thyristor. The calculation time saved is far bigger for dynamically triggered structures and can reach three decades.

Behavioral models present a clear gain in terms of calculation time. They also present the advantage of solving convergence problems, thanks to the elements of the language that allow the redefinition of the operating points of the system during the triggering of protections.

4.3.6. Addition of failure criteria

Users also expect a model to be able to predict a system failure. We therefore added a robustness criterion that allows prediction of the amplitude and time of the structure or function failure from a dynamic simulation. The prediction of material and/or functional malfunctions is possible thanks to the addition of failure criteria obtained through measurement.

4.3.6.1. Material failure

According to the studies by Wunsch and Bell [WUN 68], the thermal destruction of semiconductors is strongly linked to the amplitude and length of the pulses. The analytical model developed by Wunsch and Bell shows that destruction depends strongly on the parameter $\frac{1}{\sqrt{t}}$, where t is the length of the pulse. This formulation has been extended by Tasca [TAS 70] to adiabatic regimes with a form in $\frac{1}{t}$.

To build our behavioral mode, we reused the equations from Wunsch and Bell and from Tasca. They were simplified as much as possible to have as few parameters as possible. The equation implemented in the body of the VHDL model is the following:

$$Pf = \frac{A}{t} + \frac{B}{\sqrt{t}} + C \qquad [4.18]$$

where P_f is the power to failure and A, B, and C are empirical parameters used to correlate with the measurements. An example of a curve obtained through measurements by modifying the length of the TLP pulse for a snapback structure is shown in Figure 4.47. The failure points I_{T2} obtained for different TLP pulse lengths provide the blue curve by interpolation using equation [4.18]. The simulations carried out with the behavioral model that we developed are also shown.

During the temporal simulation, the energy dissipated into the structure is calculated using the following formula [4.19] in order to predict the failure. The temporal energy is compared to the Wunsch and Bell curve:

$$E = \int i(t).v(t)dt \qquad [4.19]$$

In the case of ESD protection structures, the value of the internal resistance is very small. As a first approximation, we consider that the voltage between their terminals does not vary. As soon as the structure is triggered, the running time (t_0) is saved. From there, the evolution of $\int i(t-t_0)dt$ is compared to $f(t-t_0)$, $f(t)$ being the equation corresponding to the measured Wunsch and Bell curve [4.18].

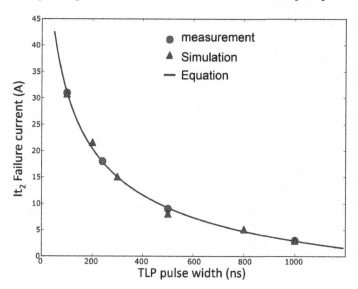

Figure 4.47. *Wunsch and Bell curve representing the maximal TLP current withstood by the structure as a function of the width of the TLP pulses, measured in the LIN component. For a color version of this figure, see www.iste.co.uk/bafleur/esd.zip*

As soon as $\int i(t{-}t_0)\ dt \geq f(t{-}t_0)*t$, the component is identified as failing, which means considering the energy received to be greater than the energy determined by the Wunsch and Bell curve.

This technique was used to analyze the failure of a LIN component with and without external capacitance (Chapter 5 – case study 5). Table 4.8 compares the robustness tests carried out on the LIN with the simulation, according to the requirements of the HMM test (where the ground of the ESD gun is directly linked to the electronic board). According to the results, we are able to predict the level of robustness with relatively good precision, with or without external components.

The combination of behavioral models extracted from TLP measurements and the extraction of Wunsch and Bell curves is therefore a good analysis element to predict the robustness of components. The method is obviously subject to a good transient simulation.

Test HMM 150 pF	Measured robustness	Simulated robustness
Without external capacitance	5.5 kV	5.8 kV
With a capacitance of 10 nF in parallel	12 kV	11.5 kV
With a capacitance of 100 nF in parallel	>25 kV	>25 kV

Table 4.8. *Comparison between the robustness tests of a LIN component according to the recommendations of the HMM, measured and simulated. The measurement conditions are limited to 25 kV*

4.3.6.2. Functional failure

Functional failures are very diverse. They depend on the application and the degree of safety required for it. For example, the critical real-time onboard systems used in automotive and aeronautical applications cannot undergo any clock cycle losses, which could cause a system response delay and be detrimental to passenger safety as a result. However, for audiovisual applications, the long response times (several ms) are not a critical issue. The definition of the failure criteria of systems depends on the applications. Some of the most common are the RESET, memory soft failures, clock time loss for digital applications and, for analog applications, uncontrolled transient voltage or current levels.

Given these remarks, it seems clear that the precision obtained during transient simulations is the most important criterion when looking to deal with functional failures. This is illustrated in Chapter 5 through two case studies:

– impact of a decoupling capacitance on the clock losses of a D latch setup as a 2-divider – section 5.3.4;

– susceptibility of a 16-bit microcontroller dedicated to automotive applications undergoing an ESD stress: set to RESET after loss of several clock periods – section 5.5.

4.4. Conclusion

Modeling is an essential tool for designing effective ESD protection strategies, as much at the level of the integrated circuit as for the board system.

Numerous research projects have led to the development of the methodologies that have proven to be effective.

At the level of the integrated circuit, the evolution of Computer Aided Design (CAD) for use in ESD modeling is relatively slow. The physical simulation tools (TCAD) marked the first revolution, followed by compact modeling to simulate the association of protections and verify the global protection strategy at the chip level. Currently, given the complexity of components, the verification tools of the global ESD protection strategy of a chip (ESD EDA checker) are developed internally by IC manufacturers [LES 15, GEV 15, ETH 15].

At the system level, there are not really any CAD tools available and the challenge is to convince IC manufacturers to complete the IBIS models with models that are specific for the ESD protections. A working group from the ESD Association, WG26 – *Models for system level simulation*, studies the use and the formalism of behavioral models, allowing exchange of models while guaranteeing intellectual property.

5

Case Studies

In this chapter, we describe several case studies that have been analyzed at LAAS-CNRS within the context of PhD theses and with our industrial partners. These concern both integrated circuits and electronic systems on boards.

5.1. Case 1: Interaction between two types of protection

This case study is concentrated around the analysis of the failure of a test vehicle intended for the development and the optimization of an ESD protection strategy for an analog 1.2 μm CMOS technology [TRE 04a]. For this, the test vehicle used is composed of an inverter circuit with an input IN, an output OUT and a power supply (VDD and VSS). As is shown in Figure 5.1, this inverter equally consists of ESD protection for the input composed of a double protection in π made of two NMOS transistors (GCNMOS) M1 and M3, a resistor, a diode D2, and a PMOS M4 transistor. The central protection between the VDD and VSS power supplies is a NMOS M2 transistor with a gate capacitive coupling (GCNMOS). These different protection components have been independently optimized and follow design rules that provide them with an HBM robustness that is equal to or greater than 6 kV.

Before its implementation on silicon, the protection circuit was first simulated and provided efficient protection for the different stress configurations. The target protection is 4 kV HBM. The test circuit was then characterized by an HBM tester for different pin combinations, and the results are presented in Table 5.1. Contrary to expectations, although each protection has an HBM robustness superior to the 4 kV targeted, one of the combinations, between IN and VDD, does not reach the specification. The robustness is only 3 kV.

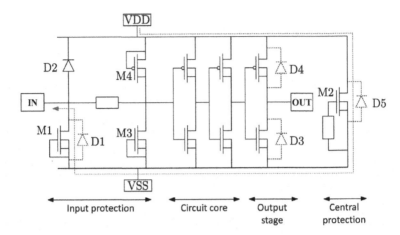

Figure 5.1. *Electrical diagram of an ESD test vehicle*

HBM (kV)>0	IN	VDD	OUT	VSS
IN		6	5	13
VDD	3	7	5	
OUT	6.5	7.5		15
VSS	14		16	16

Table 5.1. *Results of HBM tests between each pin. A pin from the first column is positively stressed with respect to a pin from the first row*

In order to understand the origin of this premature failure, a non-destructive failure analysis was carried out. The electrical signature of failure is a leakage current between VDD and VSS. To locate the origin of the leakage, the EMMI technique was first used, but without success. However, an OBIRCH analysis allowed localization of the default, as shown in Figure 5.2. It is located in the latch-up ring of the protection transistor M1. This default permits the assumption that the parasitic NPN transistor, formed by the drain of transistor M1 (emitter), the substrate (base) and the N^+ protection ring against latch-up (collector), was activated during discharge.

Protection rings against the latch-up protect the integrated circuit from the triggering of a parasitic thyristor that can lead to the destruction of the circuit. It consists of diffusions surrounding the active components that are connected either to the power supply or to the ground. Under standard operating conditions, they are therefore biased and collect potential carriers injected by the active component in the substrate. However, in the case of an ESD test, according to the chosen pin configuration, these rings can get in the way of the discharge. This is the case here where two parasitic transistors are formed, Q1 and Q3, as shown in the electrical diagram Figure 5.3.

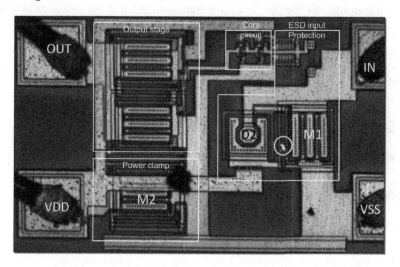

Figure 5.2. *Location of the failure in the circuit (circle) by OBIRCH*

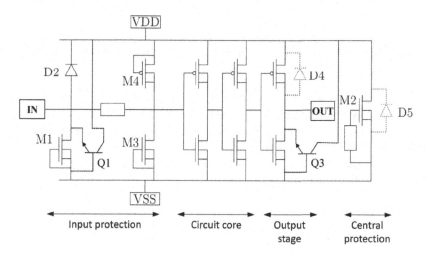

Figure 5.3. *Electrical diagram of the ESD test vehicle with its parasite bipolar components Q1 and Q3 associated with the latch-up rings*

Figure 5.4 shows a magnified view of the defective zone and displays the parasitic component responsible. Indeed, during a positive discharge between IN and VDD, this parasitic NPN transistor is activated as diode D1 (drain-substrate diode of GGNMOS M1) marked on the diagram of Figure 5.1, conducts in forward mode and acts as the emitter for the parasitic NPN transistor. As it is not designed to absorb high currents, this leads to a thermal runaway and a localized fusion of silicon.

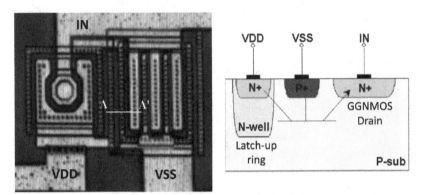

Figure 5.4. *Magnified view of the location of the failure and a cross-sectional diagram A-A' of the parasitic bipolar NPN transistor associated with the latch-up ring. For a color version of this figure, see www.iste.co.uk/bafleur/esd.zip*

In order to validate our hypothesis, we carried out SPICE electrical simulations by adding bipolar transistors in the model and the results confronted with EMMI observations.

For a TLP current of 370 mA, as shown in Figure 5.5, the current flows primarily through the central protection M2. The active path therefore goes through D1, M2, and the drain/substrate diode D3 at the output stage. The emission of the two forward-biased diodes is not visible on the figure because it is extremely weak in relation to transistor M2, which operates in the bipolar mode with its base–collector junction being reverse-biased.

A TLP current of 480 mA must be reached before the M1 protection can conduct, as shown in Figure 5.6. The current no longer flows through the central protection M2.

For a current greater than 2.3 A, the current always flows through the input protection M1, but there is an emission at the latch-up ring, which is also shown by the simulation, as shown in Figure 5.7. There is therefore another discharge path across Q1, the parasitic transistor formed by the latch-up ring with diode D1. As the latter is not optimized to conduct a high level of current, the destruction of the circuit occurs for a level of current that is far lower than the level that protection M1 can dissipate.

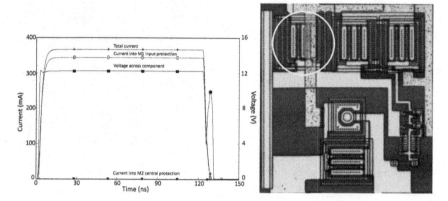

Figure 5.5. *EMMI simulation and observation for a TLP current of 370 mA. For a color version of this figure, see www.iste.co.uk/bafleur/esd.zip*

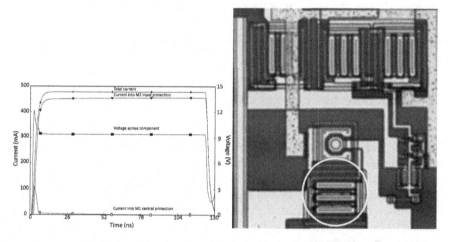

Figure 5.6. *EMMI simulation and observation for a TLP current of 480 mA. For a color version of this figure, see www.iste.co.uk/bafleur/esd.zip*

In order to avoid this premature failure, a corrective action can consist of modifying the design rules of this latch-up ring by inserting a ballast resistor (increasing the size of the ring) [TRE 04b]. This minor modification of the layout improves the robustness of the circuit while conforming to the initial predictions.

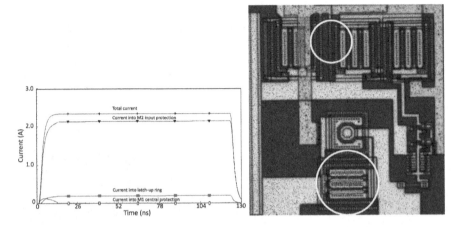

Figure 5.7. *EMMI simulation and observation for a TLP current of 2.3 A. For a color version of this figure, see www.iste.co.uk/bafleur/esd.zip*

This simple example highlights the importance of having a global protection approach in order to avoid negative interferences between different types of protection. It also shows the importance of modeling the entirety of the protection system without forgetting parasitic components. One of the challenges of circuit design tools concerns the automated retrieval of potential parasite components and their modeling.

5.2. Case 2: Detection of latent defaults caused by CDM stress

The reliability levels required by the majority of applications are extremely high. For some, such as automobiles or aeronautics, the requirement is close to zero defaults for a determined lifespan. In this context, robustness to electrostatic discharges constitutes a real challenge for new technological generations following the drastic shrinking of the size of their features. In particular, CDM discharges arising from the discharge of a previously charged component exhibit certain characteristics such as a short rise time (<0.5 ns), high peak currents (several A) and a more complex discharge mechanism. Indeed, the presence of charges in the entire volume of the circuit requires an appropriate discharge path, which is not always considered by protection strategies addressing the ESD HBM standard test. When the ESD protection is not effective, defaults may be generated within the circuit core, generally as oxide degradation, which is particularly hard to detect in a complex circuit. These defaults, which can be qualified as latent when they do not alter the operation, can nonetheless reduce the long-term reliability of the circuit concerned.

During ESD qualification, after each stress, the integrated circuit must preserve its functionality and its nominal leakage current must stay below a level defined by the specifications. However, a slight increase in the leakage current is also proof that the circuit has undergone physical deterioration. It is a latent default that can influence the operation of an integrated circuit and/or accelerate its aging. CDM stress is likely to induce defaults in the oxide, for example, trapped charges in the gate oxide that are generally difficult to detect through a measurement of the leakage current I_{ddq} (hundreds of nano-amperes) in a complex circuit.

In order to better understand both the phenomenon and the criticality of latent defaults induced by a CDM-type ESD stress, as well as the real impact of these defaults on the reliability of microelectronic circuits, the measurement technique of low frequency noise was used to study latent defaults in a commercial DC-DC converter, following a CDM stress [GAO 09].

The component studied is a boost DC-DC converter (Figure 5.8) with a pulse width modulation (PWM) that works at a frequency of 1 MHz. Regulation is optimized for applications with a constant current, such as LEDs (light emitting diode) in portable equipment, such as cell phones, digital cameras, MP3 devices, etc. Its maximal power capacity is 500 mW for 2–5 LEDs connected in series. The boost converter works in two phases; first, the inductance is charged by battery power and the energy is stored, following this, the energy is then transferred to the LEDs through an internal inverter. An external capacitor C_{OUT} (not shown in the diagram) is used to store the energy during inductance discharge (also external) and to provide current to the LEDs during the charge of the inductance.

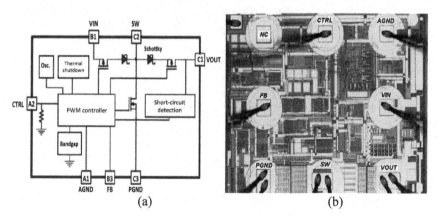

(a)   (b)

Figure 5.8. *Studied DC-DC converter: operating diagram (a) and a view of the chip mounted on a DIL24 housing (b). For a color version of this figure, see www.iste.co.uk/bafleur/esd.zip*

Input voltage VIN can vary from 2.7 V to 5.5 V according to the power needs of the different LED configurations. An external resistor R_{FB} is used to define the level of the current circulating in the LEDs ($R_{FB}=V_{FB}/I_{OUT}$) and to sense feedback voltage (V_{FB}), usually regulated at 500 mV during normal operation. The CTRL pin can be controlled by a low frequency PWM (100 Hz–1 kHz) by modifying the duty cycle to change the output current I_{OUT} with the LEDs' luminosity. A resistor of 250 kΩ is connected from the CTRL pin and the ground in order to bias that pin to 0 V while it is floating. Thus, the circuit is protected against short-circuiting, over-voltage, over-heating and electrostatic discharges.

This DC–DC converter has been optimized for an ESD robustness of 2 kV HBM and 200 V MM, but neither a test nor even a specific design has been carried out for a CDM stress. The ESD protection strategy for this component is illustrated in Figure 5.9. A BIGMOS controlled by an RC trigger circuit is used between the VIN and AGND pins. An NPN bipolar transistor is added between CTRL and AGND pins. Pins FB and VOUT are connected to PGND by the bidirectional SCR protection structure (known as "back to back"). The different grounds, AGND and PGND, are connected by two back-to-back diodes. An LDMOS transistor connected directly between the SW and PGND pins makes up the output stage of the power of the component, this pin is self-protected and no other protection structure is added.

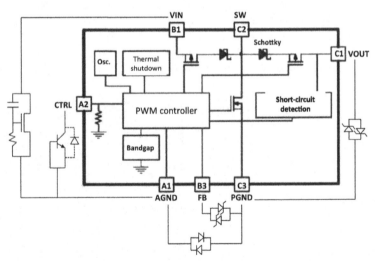

Figure 5.9. Protection strategy for the studied DC-DC converter. For a color version of this figure, see www.iste.co.uk/bafleur/esd.zip

The technology used for the creation of these circuits is a 0.8 μm BiCMOS technology which is a technology on a lightly doped (10^{15} cm^{-3}) P-type substrate, including a buried layer (N) for isolation. The typical thickness of the gate oxide is 20 nm and its static breakdown voltage is of 18 V. However, this voltage, measured in pulse mode, increases to 30 V for a TLP stress of 100 ns with a rise time of 2 ns. It increases to 43 V for a VF-TLP stress of 5 ns with a rise time of 300 ps.

For the requirements of this study, we have carried out the experimental methodology outlined in Figure 5.10 in which we have used, alongside classical techniques, low frequency noise measurement.

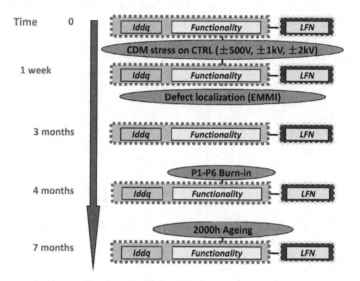

Figure 5.10. *Experimental methodology put in place for monitoring latent defaults induced by CDM stress. For a color version of this figure, see www.iste.co.uk/bafleur/esd.zip*

For this design of experiment, 27 samples of the circuit were used, of which one was kept as a reference. Twenty-four samples were organized into four test groups in order to apply FICDM stress (Field-Induced CDM) of ±500 V at ±2 kV on different critical pins (CTRL, SW, VIN and AGND). In order to limit the number of defaults, and to avoid a cumulative effect, each sample is tested on only a single pin and for a single level of CDM stress. Two samples were nonetheless reserved for the study of the cumulative effect and the impact of the stress polarity.

Prior to the application of stress, all the components were measured in low frequency noise, indicating a level of LF noise close to the minimum amount measurable at 2.3 10^{-24} A^2/Hz, staying almost entirely constant in relation to the frequency. The functionality test was also carried out, showing a very low leakage current I_{ddq} of 0.27 µA when VIN=4.2 V is in stand-by mode (CTRL is connected to the ground in order to deactivate all circuit functions). The same tests were carried out directly after CDM stress was applied.

Figure 5.11 presents the results of I_{ddq} quiescent current measurements just after CDM stress on all 24 components that had undergone a single zap on a single pin and on the two samples stressed to 500 V in a cumulative manner (five successive zaps). We observe that all the components exhibit an increase in their leakage current. A first analysis concluded that negative CDM stresses create a far greater degradation of the leakage current than positive stress. This tendency is equally confirmed by the sample stressed in a cumulative manner.

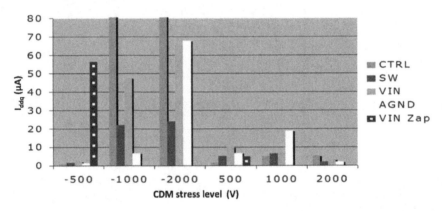

Figure 5.11. *Measurement of the quiescent current Iddq@4V on pin VIN in stand-by mode after the application of CDM stress for the 24 samples having been exposed to a single zap on a single pin and the two samples stressed in a cumulative manner to ±500 V (VIN Zap) on pin VIN. For a color version of this figure, see www.iste.co.uk/bafleur/esd.zip*

Table 5.2 summarizes the results of the complete experimental design for six samples and the reference component. We observe that all the components are considered as defective directly after the CDM test but after 3 months of storage, and even after annealing and aging, the two components P1 and P2 have returned to normal values and as such are considered to be good.

An analysis assisted by emission microscopy (EMMI) has permitted the localization of a default in one of the transistors from the start-up block of the circuit at the input of the CTRL pin (Figure 5.12).

Piece No.	P1	P2	P3	P4	P5	P6	REF
V$_{CDM}$ on pin CTRL (V)	+500	−500	+1k	−1k	+2k	−2k	*NA*
I$_{ddq}$@4.2V after stress (μA)	1.3	0,75	5.1	390	5.3	430	*0.27*
Status after CDM stress	*F*	*F*	*F*	*F*	*F*	*F*	*OK*
I$_{ddq}$@4.2V after 3 months (μA)	0.76	0,62	3.9	1.0	3.7	438	*0.27*
Status after 3 months	*OK*	*OK*	*F*	*F*	*F*	*F*	*OK*
I$_{ddq}$@4.2V after annealing (μA)	0.2	0,2	0.56	57	1.2	91	*0.37*
Status after annealing	*OK*	*OK*	*NA*	*F*	*F*	*F*	*NA*
I$_{ddq}$@4.2V after aging (μA)	0.4	0,27	NA	NA	NA	NA	*0.25*
Status after aging	*OK*	*OK*	*NA*	*NA*	*NA*	*NA*	*NA*

Table 5.2. *Evolution of the leakage current I$_{ddq}$ on VIN throughout the experimental design for six samples (P1 to P6) stressed on the CTRL pin, REF being the reference component. The status row indicates the results after the functionality test with "OK" for a successful test, "F" for a failure and "NA" when not applicable*

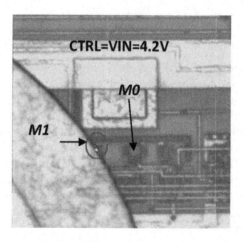

Figure 5.12. *Failure analysis by emission microscopy of the component P6 showing a light emission on transistor M1 and the input circuit. For a color version of this figure, see www.iste.co.uk/bafleur/esd.zip*

In order to verify that components P1 and P2 have truly returned to their original characteristics, we carried out noise measurements and extracted the different noise components using a simplified model (equation [2.26]). The spectral density of the noise of component P1 was primarily dominated by a G-R noise source with a cutoff frequency of 7 kHz. As for component P2, a G-R noise source with a cutoff frequency of 10 kHz combined with a $1/f$ noise source provides the best match between the measurements and the model. This match is best reached for part P3 through a combination of different G-R noise sources, and $1/f$ noise.

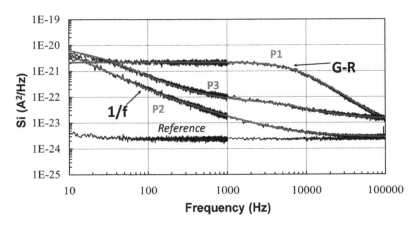

Figure 5.13. *Low frequency noise measurement after CDM stress (CTRL=0 V, I_{ddq}=270nA@4,2V before stress and I_{ddq}=600nA@4,2V after stress). The solid red lines correspond to the model (see equation [2.1])*

For sample P1, the applied CDM stress is positive with a voltage of +500 V. We propose the hypothesis that the oxides were exposed to strong electrical fields. A defect in the gate oxides of one or many of the MOS transistors at the input stage is probably created. This oxide defect can either be trapped charges in the oxide or a conduction path across the oxide. It could be the cause of the low increase in the leakage current I_{ddq} that was measured, as well as the additional G-R noise source that was observed. These charges, normally positive, can be untrapped by thermal excitation and lead to a decrease in the leakage current. This behavior is confirmed by the measurements of current I_{ddq} after thermal storage and annealing as the leakage current decreases. After annealing, this value passes below that of the reference component.

To verify that component P1 has indeed returned to its initial characteristics, we also carried out noise measurements after storage, annealing and aging. The entirety of these measurements is shown in Figure 5.14. The most remarkable result is that even following annealing, the component that shows a lower leakage current than before stress is still not returned to its initial noise characteristic (Figure 5.14(a)). Following aging (Figure 5.14(b)), we can observe that the noise level increases.

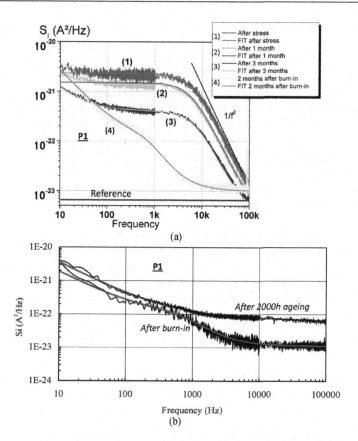

Figure 5.14. *Measurements of low frequency noise on the P1 component after a CDM stress of +500 V and annealing (a) and after aging of 2000 h at 85 °C under normal operation with VIN=3.6 V (b). For a color version of this figure, see www.iste.co.uk/bafleur/esd.zip*

We can thus conclude that the default caused by CDM stress was not only charge trapping but probably also a latent default associated with a junction and/or the Si/SiO$_2$ interface, probably microfilaments, as can be supposed from the evolution after aging.

This study shows that a CDM stress can cause a latent default in a component that is likely to evolve with time, this being a default that has not been detected by classical functionality tests.

An in-depth analysis of the protection strategy and its efficiency in relation to CDM stress has highlighted that the former was not triggered fast enough and as such resulted in the generation of overvoltages beyond the breakdown voltages of the gate oxides [GAO 07].

5.3. Case 3: The impact of decoupling capacitors in propagation paths in a circuit

The study is based upon Texas Instrument's logic inverter circuit, reference SN74LVC04A, in 0.25 μm CMOS technology and 14-pin SOIC package. This case study [MON 13b] aims to study the propagation of currents coming from ESD events (TLP pulse) and sent directly onto a system connector. The impact of all the elements that make up the system (tracks on the board, passive elements, the control box, the ESD protection strategy and ESD protections) that can impact the current waveform is studied and interpreted. Through this example, we will demonstrate the relevance of injection methods and of *in situ* voltage and current measurement techniques presented in section 2.2.

The simplified diagram of the test configurations of the system is shown in Figure 5.15. The system is made up of an inverter, some passive elements (resistor and capacitor), tracks on the board and the test environment (ESD generator, cables and probes). The circuit is not powered in this test. The injection is carried out with the TLP pulse generator. Two test configurations are studied:

– Test configuration no. 1: a study on the impact of the external capacitor C1, used as a decoupling capacitor for the circuit's power supply, when a TLP pulse is applied to the output (OUT).

– Test configuration no. 2: a study on the impact of the external capacitor C2, connected between the input (IN) and the ground plane. The TLP pulse is applied to the input (IN). This case study is detailed in section 4.3.

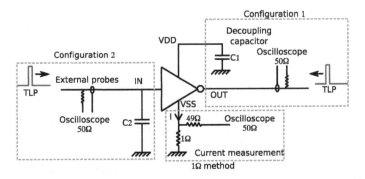

Figure 5.15. *Simplified diagram of inverter test configurations 1 and 2*

Following identification of the integrated ESD protections by TLP measurement, we created the model of the circuit's protection strategy, whose diagram is shown in Figure 5.16. This is composed of a thyristor at the input, two diodes at the output and an active PowerClamp between the power supply pins. To compare the differences in terms of simulation performance of the models, the protections were modeled by following two different methods: compact and behavioral.

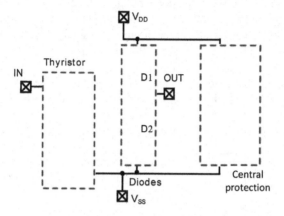

Figure 5.16. *Simplified diagram of inverter test configurations 1 and 2*

5.3.1. Analysis of test configuration no. 1

By using the system simulation method outlined in section 4.2, the simulation diagram presented in Figure 5.17 allows us to predict the paths followed by the current in the component by following the protection strategy shown in Figure 5.16. As a reminder, the blocks called "simple line" are VHDL-AMS models which allow simulation of the propagation phenomena in board tracks, whereas blocks called "coupled lines" help describe couplings between near tracks.

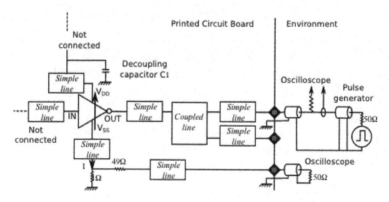

Figure 5.17. *System model and integrating circuit 74LVC04A for test configuration no. 1*

The TLP generator, connected to the output (OUT), is configured to generate a pulse of 1 A, with a width of 100 ns, and a transition time of 1 ns.

The simulation result, when the decoupling capacitor is not connected, is shown in Figure 5.18. Current I corresponds to the current, which circulates in the 1 Ω resistor. Vi is the internal voltage of the circuit between the V_{DD} and V_{SS} rails, that is to say the centralized protection pins (PowerClamp – PC). The entirety of the current injected into the output OUT (pulse of 1 A for 100 ns) circulates in the 1 Ω resistor. As expected by the circuit's protection strategy (Figure 5.19), the current is deviated by diode D1 and the PC in reverse mode in order to be evacuated by pin V_{SS}. The voltage that appears in Vi (3.8 V) corresponds to the voltage drop across the PC's pins due to its I(V) characteristic for a current of 1 A.

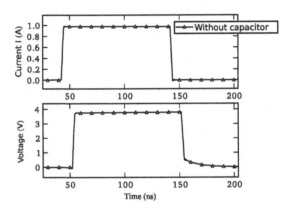

Figure 5.18. *Simulations of current I circulating in the resistor of 1 Ω and internal voltage Vi for an injection of 1 A, 100 ns, between the output and the ground plane, without a decoupling capacitor*

When a decoupling capacitor of 50 nF is connected, the waveform of current I circulating in the 1 Ω resistor is completely modified, as is shown in Figure 5.19. Complex interactions appear between the chip, its package, the board tracks and the decoupling capacitor.

In the initial state, the C1 capacitor is discharged, and the Vi voltage is worth 0. When TLP is injected into the output, the current is deviated by diode D1 as expected. At the node V_{DD}, the C1 capacitor, which is connected externally, offers a path with less impedance than the centralized protection, creating a preferential path for the discharge current.

The observed peak for current I on the resistor at time t1 must be explained. During the rapid rising edge of TLP, between t0 and t1 (Figure 5.19), the inductive effect of the package (inductance of 3.9 nH) will allow the PC to conduct a part of the current for a very short lapse of time. Indeed, the inductance of the package is not able to absorb the current instantly due the basic equation V=L dI/dt. As a result, the voltage at the terminals of the inductance of the package pin V_{DD} will greatly increase. The trigger voltage of the PowerClamp is reached, allowing conduction of the current. Figure 5.20 shows the path of the current in the circuit during the TLP rising transition.

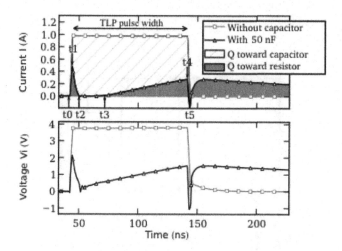

Figure 5.19. *Simulations of current I circulating in the resistor of 1 Ω and internal voltage Vi for an injection of 1 A, 100 ns, between the output and the ground plane, without a decoupling capacitor and with a capacitor of 50 nF*

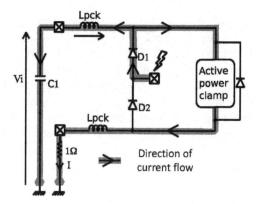

Figure 5.20. *Simplified diagram of the circuit showing the path of the current during TLP rising transition, with a decoupling capacitor of 50 nF, between t0 and t1*

Following the rapid TLP rising phase, at t1+ε, the current becomes constant. The inductive effect of the package decreases as does the voltage at its terminals. The PC's trigger voltage drops and it deactivates itself.

Between the intervals t2 and t3, all injected current leaves the circuit through pin V_{DD}. The external capacitor (C1) begins to charge. No current flows across the 1 Ω resistor. When the charge voltage of the capacitor reaches the trigger voltage of the PC at t3, the PC is triggered and conducts more and more current as the capacitor charge increases. The voltage across the terminals of the PC defines the quantity of current it can draw.

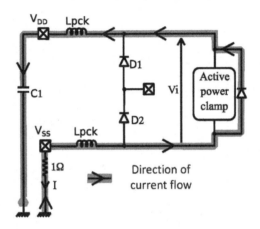

Figure 5.21. *Simplified diagram of the circuit showing the path of the current during a TLP falling transition, with a decoupling capacitor of 50 nF, at time t4*

When the TLP pulse is ended, we are confronted with a rapidly decreasing transition, at time t4. A negative peak is visible on the simulation results (Figure 5.19). This can be explained by what is called "the effect of balancing the inductance of the package". The path of the current in the circuit during the negative TLP transition is shown in Figure 5.21. Just before the end of the TLP pulse, at t4-ε, the current circulates primarily in the inductance of pin V_{DD} (approximately 700 mA for 1 A injected) and charges the external capacitor. When the flow of the injected current stops, the current cannot instantly vary in the inductances. Current continues to circulate in the inductance of pin V_{DD}, forcing the PowerClamp to conduct in forward mode in order to respond to the effect of balancing the inductances. The current flow reverses and circulates from the ground toward the capacitor.

After the pulse, at time t5+ε, the decoupling capacitor discharges across the PC until the latter switches off. Even if the decoupling capacitor absorbs a large proportion of the energy corresponding to the hatched area in Figure 5.19, the PC must conduct over a longer time period than the duration of the stress in order to dissipate the charges accumulated in the capacitor. Even though the current that circulates across the PC is weaker than the magnitude of the injected current, the latter continues to conduct for a relatively long time after the end of the discharge, which can cause the protection to leave its SOA zone.

Once the PC is deactivated, the remaining charges that have been accumulated by the capacitor are blocked and result in a high voltage between V_{DD} and V_{SS}. This can have a crucial effect on the susceptibility when the circuit is functioning.

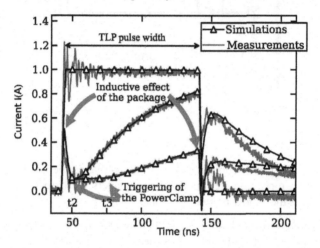

Figure 5.22. *Comparison of measurements and simulations of the current circulating in the resistor of 1 Ω, for and injection of 1 A between the output and the ground plane, with and without the decoupling capacitor of 14 nF and 50 nF*

The comparison of the simulation results with 1 Ω measurement is laid out in Figure 5.22. For the different values of the capacitor (50 nF, 14 nF and without capacitor – the value 14 nF was chosen in order to best visualize charge phenomena), the simulations correlate perfectly with the 1 Ω measurements. Initially, this current was not predicted in the previous simulations. It corresponds to the conduction of the NMOS PullDown transistor of the circuit output buffer. During discharge, the rising transition of the TLP pulse biases the gate of the NMOS transistor by capacitive coupling (drain/gate), thus allowing its conduction. The current circulating in the transistor is of 100 mA for a stress of 1 A. It cannot be predicted by only considering the ESD protections in the simulation.

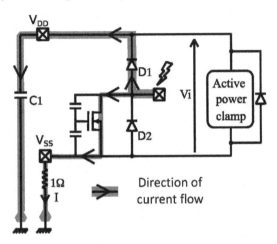

Figure 5.23. *Connection of PullDown transistor at output stage to take into account the gate coupling*

In order to take into account the current circulating in the PullDown transistor, we have added a more complex model that considers the output transistors of the circuit. We tested the following two solutions:

– First, we implemented a level 1 SPICE model that we optimized so that its characteristic fits with that given by IBIS files. This model is added, as illustrated in Figure 5.23, in parallel to the compact models of ESD protection. Capacitors, representing the gate-source and gate-drain capacitors, were added to reproduce the coupling. This model helps obtain the simulation results shown in Figure 5.24, regardless of the configuration studied.

– A purely behavioral simulation was also carried out. IBIS models of the output buffer were added. ESD protections were modeled following behavioral modeling methods. A comparison between compact and behavioral modeling including IBIS models is shown in Figure 5.24 in the case of a decoupling capacitor of 50 nF. A small difference of 50 mA is observed between the two models, but this is negligible when the magnitude of stress is increased. The level of the current circulating in the PullDown transistor is not directly linked to the ESD pulse, due to the fact that the latter is in saturated regime and thus remains limited. However, if avalanche breakdown is reached, the current increases sharply, which, in most cases, leads to destruction of the component [GIR 10].

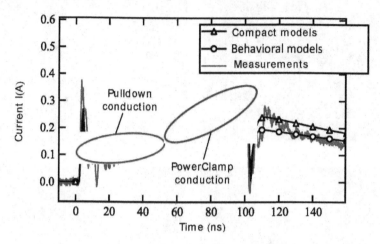

Figure 5.24. *Comparison of measurements and simulations of the current circulating in the resistor of 1 Ω, for an injection of 1 A between the output and the ground plane, with and without the decoupling capacitor of 14 nF and 50 nF for compact and behavioral modeling techniques*

5.3.2. Validation of the results by near-field mapping

Near-field mapping, as introduced in section 2.2, was used on the inverter board in order to view the distribution of the current. To recall, the mapping obtained by repetition of the pulses allows construction of the distribution of the magnetic field point by point. Through integration of the field vectors, the dynamic circulation of currents on the board can be reconstituted. Figure 5.25 shows the distribution of the current on the board during a TLP pulse of 1 A for a length of 100 ns. Each image corresponds to a different time. For the acquisition of Figure 5.25(a), obtained prior to sending the TLP pulse, no current is measured. Then, the current is injected on the OUT track, Figure 5.25(b), which is then distributed between the 1Ω resistor and the decoupling capacitor, Figure 5.25(c) to finally be dissipated by the ground, Figure 5.25(d). In the latter image, we can see some noise. This noise is linked to the return path of the current by the ground plane, which occurs between the ground via on the PCB and the ground of the connector, not visible on the image. This mapping allows the distribution of the current to be checked as explained previously.

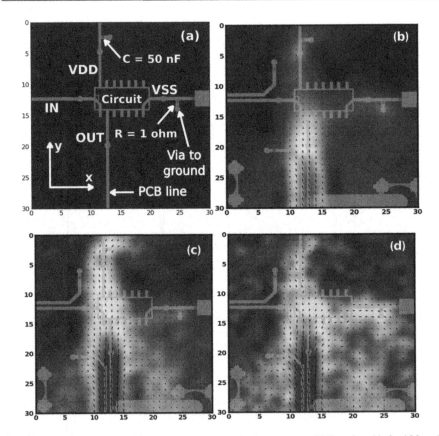

Figure 5.25. *Dynamic distribution of the currents during a TLP pulse (1 A, 100 ns) applied on the output OUT of the inverter: board before stress (a), beginning of the injection (b), charge of the capacitor C (c) and distribution of the current between capacitive path and ground (d) [CAI 15b]. For a color version of this figure, see www.iste.co.uk/bafleur/esd.zip*

5.3.3. Analysis of test configuration no. 2

In this configuration, the TLP generator is connected between the input IN and the ground plane. It is configured to send a positive pulse of a magnitude of 1.7 A and of width 100 ns with a transition time of 1 ns. We analyzed the impact of the capacitor C2 (Figure 5.15) connected in parallel to the ESD protection at the input IN (thyristor).

When the C2 capacitor is not connected, the entirety of the injected current circulates in the 1 Ω resistor. The voltage drop across the thyristor is 3 V. As expected by the circuit protection strategy, the current flows through the thyristor in the reverse mode (Figure 5.26).

When we add a capacitor of 6.8 nF between the input and the ground plane, the waveform of the current I is modified as shown in the simulation results in Figure 5.26.

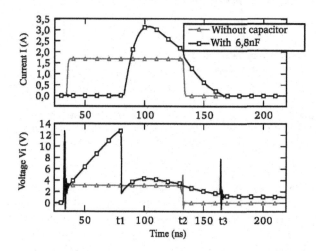

Figure 5.26. *Simulations of current I circulating in the 1 Ω resistor and the voltage Vi in the thyristor for a TLP pulse of 1.7 A injected between the input and the ground plane, with and without a capacitor C2 of 6.8 nF*

When stress is applied to the input pin, the discharge current is absorbed by the external capacitor. The thyristor is not triggered since the voltage is inferior to its trigger voltage (13 V). When this latter is reached, at time t1, the thyristor snaps back. Its voltage drops to 2 V, which is its holding voltage. The thyristor will then conduct all the injected current by the TLP, plus the current issued by the discharging capacitor. This effect of charging and discharging, associated with snapback of the thyristor, creates a high current peak of 3.2 A for 1.7 A injected. When the TLP pulse ends, at time t2, as long as the voltage of the capacitor is still higher than the holding voltage of the thyristor, it will continue to discharge across the thyristor. At time t3, the current becomes inferior to the holding voltage, the thyristor turns off, and the capacitor stays charged at around 2 V.

The simulation results compared to measurement are shown in Figure 5.27. Without a capacitor, the simulation correlates perfectly with the measurement. However, with a capacitor of 6.8 nF, a difference of approximately 25% can be observed between the simulation and the measurement.

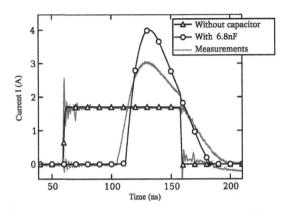

Figure 5.27. *Comparison of the measurement and simulation of the current obtained with the 1 Ω method for an injected TLP pulse of 1.7 A, with and without capacitor C2*

This difference is mainly linked with the discharge constant of the capacitor. The simulation was first carried out without taking into account the frequency models of passive components, such as the capacitor C2. In order to improve the correlation, we have included a frequency model of the capacitor C2, which takes into account inductive and resistive parasitic elements. The adjusted comparison of the measurement and the simulation is shown in Figure 5.28. The dynamics (essentially linked to parasitic elements of the capacitor in parallel to the housing) are reproduced perfectly. Both for the compact and behavioral simulations, the prediction errors over the transient are very weak. It is this optimized model that is used in the simulations in Figure 5.26.

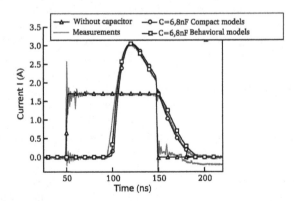

Figure 5.28. *Comparison of the measurement and simulation of a SCR trigger with and without an external capacitor: simulation of compact models (semi-empirical) and behavioral models – measurements carried out with the 1 Ω method*

These effects of the charging and discharging of the external capacitor during an ESD discharge on triggering of the ESD protection have been demonstrated by Patrice Besse in [BES 10] in automobile applications. The capacitor connected in parallel is used as an EMI (Electromagnetic Interference) filter. The same effect was also observed for an integrated capacitor on the chip and an ESD protection structure [LEE 10].

5.4. Case 4: Functional failure linked to a decoupling capacitor

Another study [MON 11b] demonstrates how a decoupling capacitor can affect the probability of generating functional failures during ESD discharges. The system is based on a D flip-flop (SN74LVC74A, 0.25 μm CMOS technology, SOIC 14 package) mounted as a frequency divider (Figure 5.29). An LDO voltage regulator (25DBVTG4) biases the circuit to 2.5 V and limits the current to 150 mA in order to avoid any latch-up problems [VOL 07]. A decoupling network composed of an inductance of 5.6 μH and a diode is inserted between the regulator and the circuit to allow for the use of the DPI injection method [IEC 06]. The stress is thus injected entirely into the component's power supply V_{DD}. A decoupling capacitor is connected between pin V_{DD} and the ground plane to study its influence on the probability of generating a functional fault. The arbitrarily chosen criterion for failure chosen is the loss of a clock edge (from the input signal CLK).

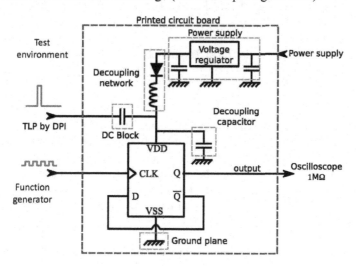

Figure 5.29. *Electrical diagram of the test board of a D flip-flop mounted as a frequency divider*

For the simulation, we followed a hierarchical description method, where the circuit is described in a behavioral manner. The simplified diagram of the model of the system used to carry out simulations in this section is shown in Figure 5.30. It includes the test environment (TLP generator, cable, probe), the board (LC line + passive components) and the integrated circuit (IBIS model for passive elements + ESD protections).

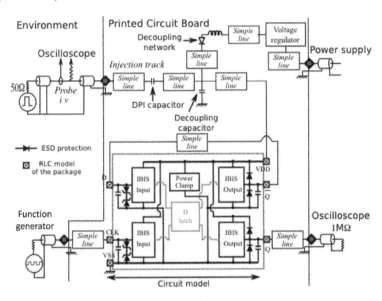

Figure 5.30. *Electrical diagram of the simulated system integrating SN74LVC74A circuit exposed to a TLP pulse by DPI. For a color version of this figure, see www.iste.co.uk/bafleur/esd.zip*

A comparison of the measurement and simulation of the system having been exposed to a pulse of 75 V TLP, with a clock frequency at input of 2 MHz, and a decoupling capacitor of 50 nF is shown in Figure 5.31. The upper curve is the input signal. As the TLP pulse cannot be synchronized, we report the observed and simulated output signal in the figure. Two different cases are observed during the experiments: when there are no losses on the clock edge, case (A) and when there are, case (B).

The failure mechanism is primarily linked to the voltage levels compatible at the input stage. Figure 5.32 shows the simplified diagram of the system (a) and the chronogram (b) supporting the explanation. During TLP pulse, the discharge current I_{TLP} charges the decoupling capacitor C. The voltage on the V_{DD} rail increases with the charge on the capacitor. After the TLP pulse, voltage V_{DD} decreases. During the charging and discharging, voltage V_{IH}, which determines the high logical level of the inputs of the circuit (Figure 5.32(a)), will vary proportionally to V_{DD}. A critical voltage (V_{crit}) is then defined when the voltage V_{IH} is greater than this voltage at the input. This level allows the critical zone illustrated in the chronogram of Figure 5.32(b) to be highlighted. If an active edge arises in this zone, with the level V_{IH} being greater than the level of the input, it will not be considered by the D flip-flop.

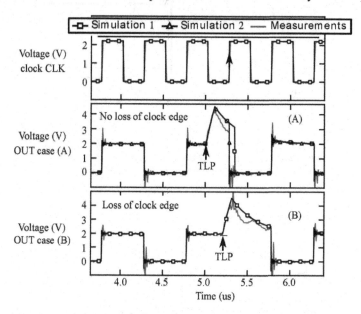

Figure 5.31. *Comparison of the measurement and simulation of the output voltage in two cases: no loss of clock edge (A) and loss of clock edge (B)*

The mechanism described appears to be perfectly normal from the point of view of signal integrity and it does not use the ESD protections. What happens when the trigger voltage of the PowerClamp between V_{DD} and V_{SS} is reached? The PowerClamp is added to the previous diagram, and its ideal characteristic is illustrated (Figure 5.33). Figure 5.33(b) shows the simulation results of the current I circulating in the PowerClamp and its voltage for a TLP injection of 400 V. The simulation is carried out with and without the PowerClamp.

The behavior is similar to the previous one until the voltage V reaches the PC's triggering voltage V_{t1}. Part of the current will then circulate across the latter, slowing down the charge of the capacitor. As a result, the PC will absorb a large part of the injected current, up to 4.5 A, according to the simulation results. The maximum voltage reached internally is limited and ranges from 16.5 V (without PC) to 11.6 V (with PC).

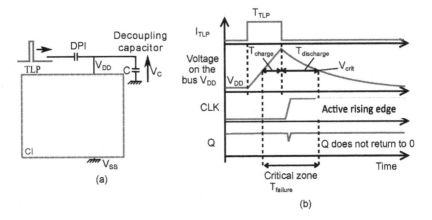

Figure 5.32. *Equivalent diagram of the circuit (a) and the chronogram showing the mechanism causing the failure (b). For a color version of this figure, see www.iste.co.uk/bafleur/esd.zip*

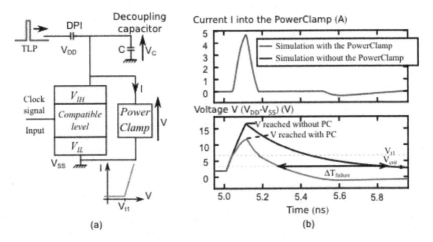

Figure 5.33. *System diagram incorporating the PowerClamp (a) and the simulation results of a TLP injection by DPI of 400 V, 8 A (b)*

Measuring the probability of generating a functional failure depends on the value of the decoupling capacitor (50 nF or 6.8 nF), on the frequency of the D flip-flop

(2.5 and 10 MHz), and on the magnitude of the TLP pulse (from 10 V to 200 V). The pulse is repeated and the number of failures is counted, allowing us to determine a percentage of error generation, as shown in Figure 5.34.

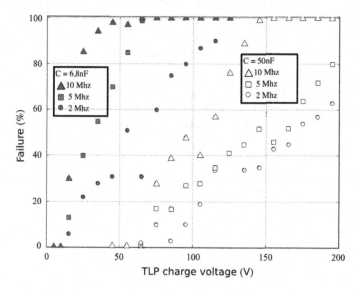

Figure 5.34. *Evolution of the probability of functional failure in relation to the value of the decoupling capacitor, to the operation frequency and to the pulse magnitude. The tendency curves have been added*

When the operation frequency of the system increases, the probability that an active rising clock edge occurs in the critical zone (see Figure 5.33(b)) also increases. Equally, when the value of the decoupling capacitor, C, decreases (from 50 nF to 6.8 nF), the probability of losing a clock edge also increases. Indeed, a lower value of C allows it to be charged faster during a TLP pulse. The voltage reached on the V_{DD} rail is therefore greater. Thus, the critical zone (Figure 5.32(b)) is bigger and the probability of failure increases.

Figure 5.35 gives a closer view of the percentage of errors obtained when a decoupling capacitor of 50 nF is connected. The added lines show the tendency of the probability. When the injection is superior to approximately 100 V, the PowerClamp is triggered and it slows down the charge of the capacitor as previously explained. As we can see from the measurement results, the tendency curve slope is reduced and two tendency curves become apparent.

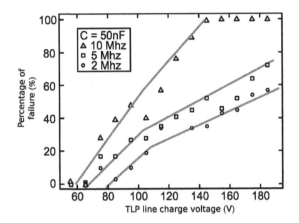

Figure 5.35. *Influence of the PowerClamp on the probability of generating a failure*

An error is generated when the voltage on the V_{DD} rail is superior to the voltage V_{crit} and if the clock edge occurs within the critical zone ($T_{failure}$), as shown in Figure 5.33. Zone $T_{failure}$ corresponds to the sum of the charge time T_{charge} and of discharge, $T_{discharge}$. By considering that the shift of the voltage on the V_{DD} rail is caused by the charge voltage of the external decoupling capacitor C, we can use the following basic equation to establish T_{charge}:

$$I = C \frac{dV}{dt} \qquad [5.1]$$

By solving the charge and discharge equations of the capacitor and by following the criteria previously elaborated, we can infer a rule on the evolution of the probability of generating a failure:

$$P_{failure} = \frac{T_{failure}}{T_{clk}} = \frac{T_{charge} + T_{discharge}}{T_{clk}} \qquad [5.2]$$

with T_{clk} the clock signal period.

Figure 5.36 reports this equation on the probability result for a capacitor value of 50 nF. As we can see, the rule is correct for weak injections. However, beyond 100 V of TLP voltage, the rule moves away from the measurements.

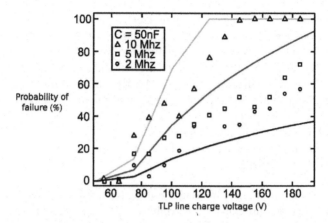

Figure 5.36. *Superimposition of the probability law with the measurement of the evolution of the probability of failure*

When the magnitude of the TLP is high enough for the voltage reached on the V_{DD} rail to be equal to or greater than the triggering voltage of the PowerClamp (around 7 V), part of the injected current flows through it. The voltage reached in the capacitor is reduced, thus diminishing the failure probability. The triggering voltage of the PowerClamp is reached for:

$$I_{TLP} = \frac{C \cdot (V_{th} - V_{DD})}{T_{TLP}}$$ [5.3]

By reintroducing this data, it is then possible to refine the laws on failure probability.

This probability approach is the only coherent way of determining if a system will have failures or not. Among the various parameters evoked, ESD protections play an important part, but the conditions for the use of the system, that is the design specifications, are just as important. A certain number of industrial partners in our research projects have already integrated the protections in their design specifications, so that they may have a reliable solution as soon as possible.

5.5. Case 5: Fatal failure in an LIN circuit

As part of the ANR project, E-SAFE [ANR 12], a consortium made up of LAAS-CNRS, Freescale Toulouse, an integrated circuit manufacturer, and VALEO Créteil, automotive suppliers, led a study on the ESD robustness of a Local Interconnect Network (LIN) component intended for communication series in automobile networks. This integrated circuit creates the physical layer of the signals, respecting communication protocol LIN 2.1. Due to rather draconian standards in the automobile industry, the circuit is optimized to pass not only standards, such as EMC DPI [IEC 07a, IEC 06], BCI (Bulk Current Injection) [IEC 07b], but also system level ESD test standards (IEC61000-4-2 [IEC 08] and ISO10605 [ISO 08]). The collaboration aimed to develop generic methods for measurement and the extraction of ESD behavioral models that would predict the impact of ESD at system level. Beyond the response of the system to a TLP pulse, we wish to predict the behavior of the system submitted to ESD gun discharges, defined by IEC61000-4-2 and ISO10605 standards.

The LIN, in an 8-pin SOIC package, is mounted upon the printed circuit board (PCB), developed and provided by the company VALEO. This printed circuit was designed to be generic and modular in such a way as to test circuits in SOIC packages with up to 14 pins. Thus, it is possible to test all types of circuits with compatible package in the same configuration [LAF 11, BES 11]. The printed circuit board integrates the different patterns for measurement and injection presented in Chapter 2. The electrical diagram of the pattern implemented on each pin of the circuit is given in Figure 5.37. This pattern is modular and allows, depending on the soldered passive elements, to carry out S parameters, EMC (by DPI injection) and ESD characterizations (TLP). All these techniques are detailed in Frédéric Lafon's thesis [LAF 11]. All the external components are mounted on a CMS package. A kit, not shown here, including the patterns for open circuit (OC), short circuit (SC) and 50 Ω, allows for the calibration of measurement instruments (the network analyzer, for example).

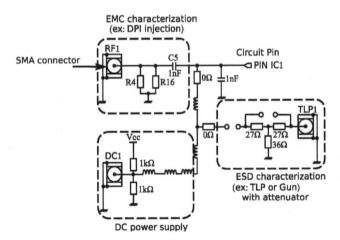

Figure 5.37. *Diagram of the injection and measurement patterns applied at each of the pins of the circuit [LAF 11, BES 11]*

The simplified diagram of the test configuration is shown in Figure 5.38. For the simulation, all the parasite elements of the package and of the PCB lines are implemented. A capacitor of 220 pF, used as an EMI (ElectroMagnetic Interference) filter, is connected to the LIN pin of the circuit in order to comply with the "LIN Conformance" [SAE 12] standards, which define OEM requirements on the LIN pin. This pin must withstand IEC61000-4-2 stresses, with and without an external capacitor of 220 pF. The following conditions were studied:

– Test with a 50 Ω TLP generator, configured to send a pulse of 5 A, with a duration of 100 ns and with a transition time of 1 ns [BÈG 15]. The measurements are carried out for the 0.1 Ω and TDR/TLP method presented in Chapter 2.

– Test with an ESD gun, pre-charged to 2 kV. In this case, only the injected current is measured with an external probe. This test configuration follows the specifications given in the HMM standard [ESD 09].

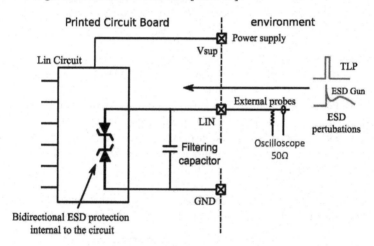

Figure 5.38. *Simplified diagram of the case study: ESD injection on the LIN pin and the study of the impact of the EMI filter capacitor*

Bidirectional ESD protection, integrated between the LIN and GND pins of the circuit, is modeled according to the methodology for behavioral modeling using a TLP measurement. A comparison between TLP measurement and simulation validates the model, as shown in Figure 5.39. The protection is made up of two snapback structures connected back-to-back.

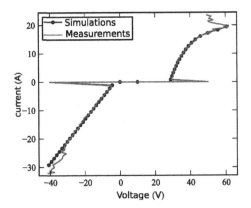

Figure 5.39. *Comparison of the TLP measurement and simulation of the ESD protection integrated between the GND and LIN pins. For a color version of this figure, see www.iste.co.uk/bafleur/esd.zip*

The board upon which the LIN component is implemented does not permit the insertion of measuring techniques close to the component. It is also difficult to perform a near-field measurement due to the four-layer structure of the board. In order to validate the models, we decided to carry out an indirect measurement with the TDR/TLP method (section 2.1.3.3). Once the system model (presented in Figure 5.38) is validated for all dynamic aspects, we expect to deduce the current circulating in the circuit through a simulation with the different configurations (with and without decoupling capacitors of 220 pF).

The transient simulations for the voltage and the current (with the capacitor of 220 pF) using the models are shown in Figures 5.40 and 5.41, respectively.

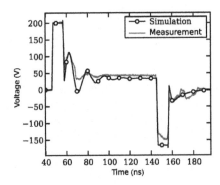

Figure 5.40. *Comparison between the measurement and the simulation of the TLP voltage waveform V(t) measured with the external probe for 200 V injected on the LIN pin, with the filter capacitor of 220 pF*

The simulation shows a good correlation with the measurement that reproduces the different peaks of the current and the voltage over time. The first part of the curves, between 50 and 60 ns, corresponds to the injected signal (200 V, 4 A). The duration of the plateau, 10 ns, corresponds to the back and forth propagation time between the measuring point and the circuit soldered to the board. At 60 ns, the large discontinuity observed is linked to the overlap of the injected pulse and the returning pulse. When the voltage reaches the triggering voltage of the ESD protection (≈45 V), the current reaches approximately 7.5 A. In less than 1 ns, the ESD structure is triggered and creates, once again, a large discontinuity, which inverses the slope of the curves I(t) and V(t). The following oscillations correspond to the R, L and C effects that are present on the discharge path of the current in the ESD protection. These elements are directly linked to the parasitic elements of the package pins, the tracks on the board and the capacitor. The first peak of the oscillation is perfectly reproduced in the simulation. The small difference observed between the simulation and the measurement (between 100 and 140 ns) is mainly due to the light error of approximation of the on-resistance value of the ESD protection.

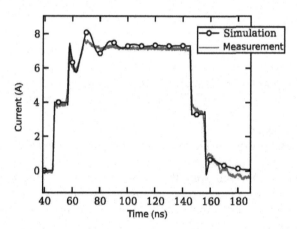

Figure 5.41. *Comparison between the measurement and the simulation of the TLP current waveform I(t) measured with the external probe for 200 V injected on the LIN pin, with the filter capacitor of 220 pF*

These measurements of returning signals assisted by TLP confirm that the models used in the simulation are correct. We can deduce the current circulating in the component. The results of the simulations, with and without the capacitor of 220 pF, are shown in Figure 5.42.

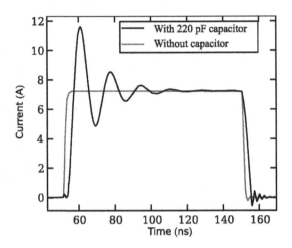

Figure 5.42. *Simulation of the current circulating in the LIN's ESD protection during a TLP pulse of 200 V, with and without an external capacitor of 220 pF*

In order to verify the robustness of the LIN component, the standard IEC 61000-4-2 is the definitive validation. The ESD guns are not 50 Ω generators. In the image of the TDR/TLP analysis that we carried out in the previous section, we studied the dynamic response of the system exposed to a discharge from the ESD gun.

The gun is charged to 2 kV. The discharge is sent into the system by direct contact to the central core of the SMA connector of the printed circuit board. The ground reference of the gun is connected in accordance with the "LIN conformance" document [SAE 12] and following the recommendations of HMM test procedures [ESD 09]. The current is measured using a magnetic probe, which is placed around the tip of the discharge gun. This permits a linear response of the current between 100 kHz and 1 GHz.

The simulation for the current and the measurement data obtained with the Fisher probe when the capacitor is of 220 pF is connected are shown in Figure 5.43.

The simulation of the injected current gives a very good image of the measured current. The issues that can be seen are essentially linked to the waveform generated by the ESD gun, complying with the standard, but that is not quite the same as the one generated by our model, which is ideal. We can confirm, therefore, that the methodology that we developed can be applied just as well for simple cases (50 Ω being adapted), as for complex cases, representative of system configurations.

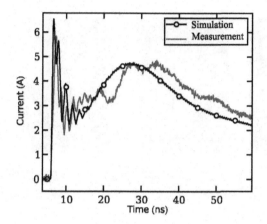

Figure 5.43. *Comparison between the measurement and the simulation of the current measured for an ESD gun injection of 2 kV (150 pF/330 Ω), with an external capacitor of 220 pF*

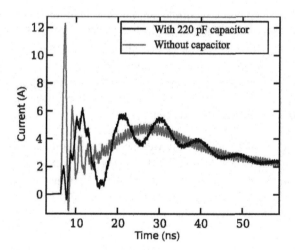

Figure 5.44. *Simulation of the current that circulates in the ESD protection of the LIN for a gun discharge of 2 kV injected on the LIN pin, with and without an external capacitor of 220 pF*

Figure 5.44 shows the results of the current circulating in the component with (in black) and without (in gray) the capacitor connected externally for an ESD gun discharge of 2 kV (150 pF/330 Ω).

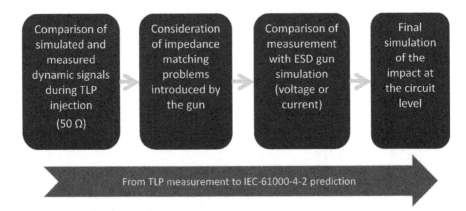

Figure 5.45. *Method for validating the setup of measurement and simulation that determines the currents across the components*

From transient simulations correlating to TLP-TDR measurements, we have validated the system model, be it for an injection with the 50 Ω TLP or for the ESD gun, following the HMM procedures and the "LIN conformance". Without any adjustments to the parameters, the simulation helped obtain a very good correlation with external measurements, showing that by rigorously applying our methodology presented in Figure 5.45, it is possible to predict the current paths that are internal to the components with relatively good accuracy.

The dynamic simulations have allowed accurate determination of the levels of robustness of the LIN component by following the methodology presented in section 4.3.4. of Chapter 4. The simulated levels are compared with the simulated levels given in Table 5.3. The results show a very good prediction of the levels of failure with and without external capacitors.

HMM 150 pF/330 Ω	Measurements	Simulations
220 pF in parallel	Pass: 15.2 kV Failed: 15.4 kV	Pass: 15.8 kV Failed: 16 kV
Without capacity	Pass: 15.2 kV Failed: 15.4 kV	Pass: 15.8 kV Failed: 16 kV

Table 5.3. *Levels of HMM robustness obtained by simulation and by measurement with and without an external capacitor of 220 pF*

5.6. Case 6: Functional failure in a 16-bit microcontroller

This last case study shows the generation of functional failures on a 16-bit microcontroller aimed for automobile application. This study was led with Freescale Semiconductor in Toulouse, in order to study the susceptibility of a 16-bit microcontroller to radiated ESD stresses [VRI 07]. In order to carry out these stresses, we used a near-field mapping method that allowed us to carry out localized injections on a circuit pin under testing. The simulations were performed at Freescale in the Cadence environment, based on the schematics of the microcontroller and on our expertise on ESD stress propagation. In this section, we start by providing some context for the study. We then describe the circuit under testing as well as the test bench used. Finally, we present the various measurements and simulations carried out.

According to the ISO 10605 standard used in the automobile field [ISO 08], a functioning electronic board must be able to withstand ESD discharges of several tens of kV in its ground plane. Measurements carried out by VALEO have permitted the observation of malfunctions appearing on an electronic board around the 16-bit microcontroller at injection levels lower than the required specifications. To determine the sensitive blocks on the microcontroller, near-field susceptibility tests are carried out by using a magnetic probe to generate locally radiated disturbances above the tested circuit. For this study, the ESD stresses were injected using an IEC

61000-4-2 generator. The malfunction criterion is the placement of the microcontroller into RESET. With this test, VALEO retrieved the susceptibility mapping for a 16-bit microcontroller submitted to radiated ESD stresses, as shown in Figure 5.46. In this figure, we can observe the required levels of magnitude needed to have a microcontroller malfunction. We note that two zones are susceptible in the circuit. These two zones are located on the RESET pins as well as the power supply pins for the oscillator and the phase-locked loop (PLL) blocks, composing the circuit's internal clock system. For the RESET signal zone, the malfunction can easily be explained. When disturbances induced by a near-field injection reaches the switching threshold of the RESET input, the RESET signal becomes activated. For the second zone, a complementary study needed to be carried out in order to understand the effects of ESD disturbances on the clock system of the microcontroller.

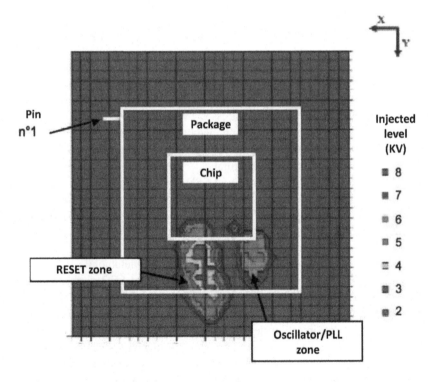

Figure 5.46. *Mapping of the susceptibility of the microcontroller submitted to radiated ESD stresses with a magnetic probe connected to an ESD gun (measurements made by VALEO). For a color version of this figure, see www.iste.co.uk/bafleur/esd.zip*

5.6.1. *Description of the studied 16-bit microcontroller circuit and its test conditions*

This circuit, designed for automotive applications, is realized in a 0.18 μm CMOS technology. The version used here is mounted on an LQFP package with 144 pins. For our study, we are interested in the part dedicated to the circuit's internal clock. To produce the clock signal necessary for the proper operation of the microcontroller, a Phase-Locked Loop (PLL) is used and synchronized on a clock reference provided by an external quartz. Figure 5.47 shows a simplified diagram of the microcontroller's generation circuit for the internal clock. This can be derived directly from the internal oscillator or from the PLL. The pins dedicated to this section are:

– V_{DDR} and V_{SS3} to power the internal voltage regulator supplying the PLL and the oscillator;

– V_{DDPLL} and V_{SSPLL} to connect a decoupling capacitor;

– EXTAL and XTAL to connect the quartz.

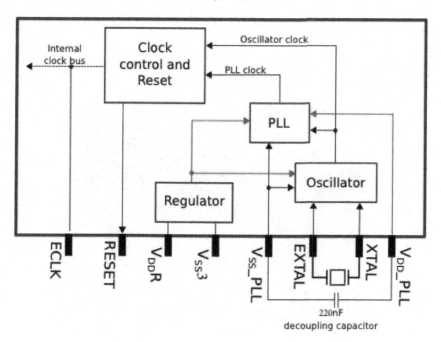

Figure 5.47. *Diagram of the principle of the circuit generating the internal clock of the microcontroller*

In order to have an image of the signal present on the internal clock rail, we monitor the ECLK pin. This not only enables viewing of the disturbances induced by the clock signal, but also acquisition of data on the modification of the clock's frequency or even its shutdown. Interruption signals can also be useful for obtaining information on the normal operation of clock circuits. In the study, however, no interruption was triggered. In order to choose the susceptibility criteria, we carried out a simple program that delivered a square signal for a period of 100 ms and was used as a reference; the susceptibility criterion is reached if the output signal remains in the off-state without automatically triggering the RESET signal. This program uses the oscillator and the PLL blocks to generate an output signal. For our experiments, we used an operation frequency of 8 MHz for the oscillator clock.

The proposed method for aggressing the microcontroller is near-field injection using a magnetic probe, as is shown in Figure 5.48. The test board is fixed on a table that can be moved in a horizontal plane along the X and Y axes. For measurement, the microcontroller is placed at the center of the board and isolated from the other components. The probe is placed 0.5 mm above the package or the circuit pins being tested. Compared to the studies carried out by VALEO using an ESD gun, a VF-TLP generator is used here to inject the ESD stress. The probe connection is made with an RG402 coaxial cable. We are able to view the control signals through 1 MΩ passive probes attached to an oscilloscope with 1 GHz bandwidth. The parameters to be taken into account for the pulses are the transients (di/dt) of the VF-TLP transitions (of a few hundred ps).

Figure 5.48. *Photograph showing a near-field injection above a microcontroller*

5.6.2. *Measurement results*

For these measurements, we viewed the following microcontroller signals:

– the ECLK signal giving the frequency of the internal bus, corresponding to the frequency of the PLL's output signal divided by two;

– the PA3 signal, called OUTPUT in our results, representing the square digital signal of a period of 100 ms;

– the RESET signal, active for a logical level of "0".

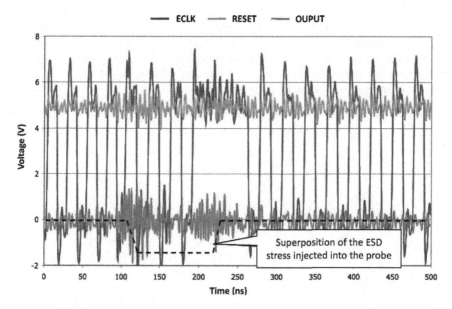

Figure 5.49. *Disturbances in the circuit's output signals without activation of the RESET for an ESD pulse of –800 V. For a color version of this figure, see www.iste.co.uk/bafleur/esd.zip*

A study on the various characteristics of the injected ESD pulses was carried out by varying the rise time (100 ps or >1 ns), the magnitude (up to 1 kV) and the polarity of the pulse (positive or negative). The zone aggressed by the probe is located between the V_{DDPLL} and V_{SSPLL} pins of the microcontroller (see Figure 5.47). Figure 5.49 shows the disturbances generated on the ECLK and OUTPUT signals for an ESD pulse of –800 V, with a duration of 100 ns and a rise time of 300 ps.

A loss of clock period on the ECLK signal is observed with the activation of the RESET signal. This causes a program shutdown, which is the most extreme case for an application, as it does not lead to restarting the microcontroller, thus resulting in function not being maintained. On the OUTPUT signal, the connection wires used for the measurement induce the observed noise. By superimposing the ESD stress sent in this measurement, two disturbance zones caused by the rising and falling edges of the pulse become visible. We can thus deduce that the rise time creates the disturbance of the PLL signal.

Figure 5.50 shows the disturbances created by an ESD pulse of –900 V on the ECLK signal. According to this figure, the frequency of PLL's output (ECLK) is no longer stable after disturbance. It is only after several clock periods that the PLL signal becomes periodic again. This is the observation on the locking of the PLL. This confirms the correct operation of this part. We therefore formulate the hypothesis that the error is generated by the oscillator block, which, for a certain amount of time, cannot deliver its output signal.

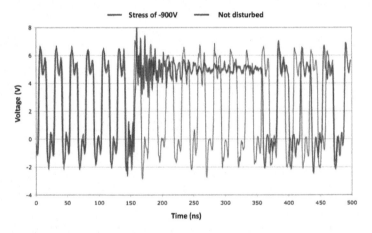

Figure 5.50. *Comparison between an ECLK signal that has not been disturbed and one disturbed by an ESD pulse of –900 V. For a color version of this figure, see www.iste.co.uk/bafleur/esd.zip*

According to the two previous figures, we can note that the more the magnitude of the ESD pulse increases, the greater the failure (loss of periods), but without the RESET signal. To activate this signal, a stress of 1,000 V magnitude must be injected. With this magnitude, the RESET signal goes to "0", which restarts the circuit and as such the PLL. Other measurement data summarized in Figure 5.51 show the influence of the rise time as well as the polarity of the injected ESD pulses.

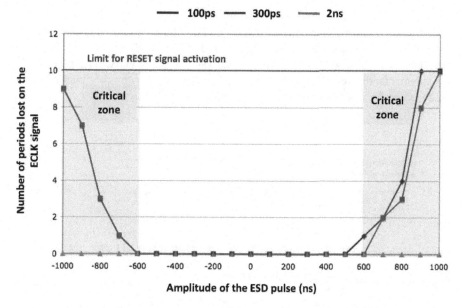

Figure 5.51. *Measurement of disturbances induced on the clock signal for different ESD pulse magnitudes and rise times. For a color version of this figure, see www.iste.co.uk/bafleur/esd.zip*

This figure shows two phenomena linked to the injection method. The first effect appears for ESD pulses from −1,000 V to −600 V and from 500 V to 1,000 V with a rise time inferior to 300 ps. In this zone, the clock signal of the circuit's internal rail (PLL's output signal) loses a few clock periods. The coupling carried out between the probe and the circuit package (the zone above the power supply block of the oscillator/PLL) disturbs the functionality of the oscillator and stops the initial program, meaning it stays in a fixed logical state and no longer delivers a square signal. In embedded applications, this failure is critical as the ESD disruption creates a system shutdown without resetting. However, for ESD injections over 1,000 V, a second phenomenon appears. With these magnitudes, the number of lost clock periods is sufficient to activate the RESET signal and as such resets the circuit. We noted that for a rise time of 2 ns, no failure is observed until a magnitude of 1 kV is reached, which is the limit of the VF-TLP generator. If we compare the results obtained with Figure 5.46 (measurement carried out with an IEC generator with a rise time of 1 ns), we note that a magnitude of 5 kV is necessary to create a circuit malfunction. This demonstrates the influence of ESD pulse transitions on the susceptibility of the microcontroller. Han *et al.* [HAN 07] carried out a similar study on an 8-bit microcontroller. The test device used for this study was an ESD injection directly into the clock system of the circuit across a capacitive probe. The measurements showed that the disruptions induced on the circuit's internal clock depended on the injected ESD stress' transitions.

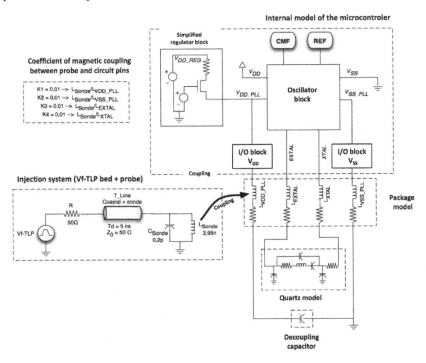

Figure 5.52. *Simplified electrical diagram of the simulation environment*

5.6.3. *Modeling and simulation*

The simulations confirmed that the clock system of the 16-bit microcontroller is disturbed by radiated ESD stress. Figure 5.52 shows the electrical diagram used to carry out this confirmation. The diagram shows the block that represents the injection system (VF-TLP + magnetic probe). To simplify the simulation models of the microcontroller, only a few of the blocks were considered:

– the oscillator block with its associated quartz, represented by an RLC model at 8 MHz;

– the models of the different V_{DDPLL}/V_{SSPLL} and EXTAL/XTAL pins;

– a simplified model of the voltage regulation block that supplies the clock system;

– the input/output blocks for VDDPLL/VSSPLL pins that allow the decoupling of the oscillator with an external capacitor.

The modeling of these different blocks was carried out using information retrieved from the circuit routing diagram. The different power supplies were modeled by ideal voltage sources. In this simulation, the clock's reference signal comes from the oscillator block and reaches directly to PLL, written REF, and the failure detection signal on the clock, written CMF ("Clock Monitor Failure") helps check the susceptibility of the oscillator block to radiated ESD.

According to the previous measurement carried out, the PLL was functioning correctly. Therefore, the oscillator could be the origin of the failure. Thanks to the model shown in Figure 5.52, we can deduce that a disturbance of the REF signal will induce a disturbance on the clock signal of the PLL, meaning the PLL block does not need to be added in this simulation. Figure 5.53 shows the simulation results of disturbances generated on ECLK clock signals for negative TLP pulses from −700 V to −1,000 V. Despite the loss of clock periods, proportional to the magnitude of the stress, no RESET was observed.

The simulations results obtained with various rise times are shown in Figure 5.54. A good correlation is obtained with measurement data shown in Figure 5.51. The few differences between the simulation and the measurement can be attributed to the inaccuracy of rise times caused by the injection device. Nonetheless, with this simulation, we can observe the disturbance induced on the circuit's internal clock signal. These simulation results demonstrate the susceptibility of the oscillator block to radiated ESD aggressions.

This study, carried out by combining the near-field probe with the VF-TLP, has allowed the local disturbance of the clock system of a 16-bit microcontroller up to malfunction (loss of clock periods). We note that an ESD pulse with a magnitude superior to 600 V and a rise time of a few ps results in the malfunction of the circuit. This is due to the pulse transitions. This has been verified using simulation with SPICE models. These simulations were only made possible thanks to the collaboration of Freescale manufacturer, who was in possession of the models. Thanks to these simulations, we identified the cause of the malfunction as coming from the loss of clock periods that does not necessarily induce the activation of the fault detection signal on the clock.

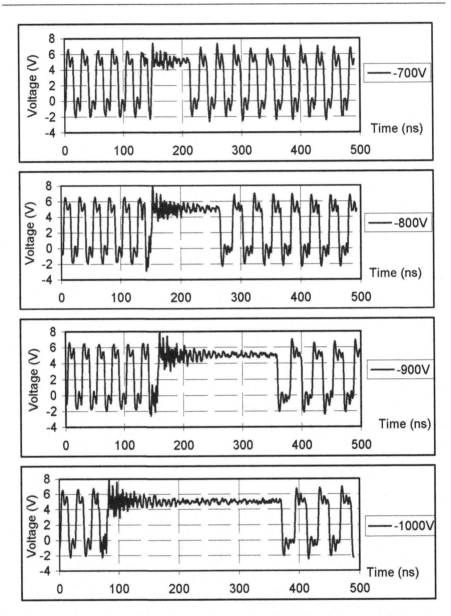

Figure 5.53. *Disturbances observed on a clock signal after applying TLP stress with a rise time of 300 ps for injections of −700 V to −1,000 V*

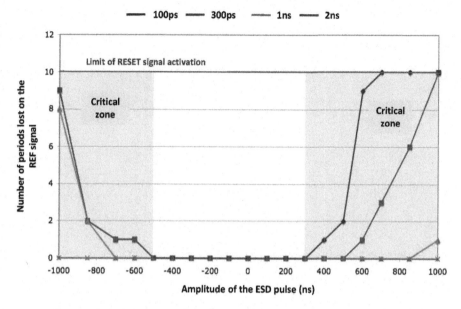

Figure 5.54. *Simulation results on the disturbance induced on a clock signal by different ESD pulse magnitudes and rise times. For a color version of this figure, see www.iste.co.uk/bafleur/esd.zip*

5.7. Conclusion

We have illustrated, using different case studies, the diversity of techniques that can be used for the optimization of the ESD protection strategy and failure analysis. They equally show the complexity associated with, and the necessity of, a global approach in order to be successful with this optimization.

In the following chapter, we attempt to summarize a set of useful and/or essential rules to achieve a protection strategy.

Conclusion

To conclude this work, we would like to recall some important rules, some of them crucial, for successfully creating an ESD protection strategy in an electronic circuit or board. The constraints – ESD, CEM, latch-up, etc. – must be considered from the moment the specifications are defined, and from the first steps of the design. This allows to define choices in terms of the architecture, adding new constraints to the functionality of this system (e.g. parasitic capacitance linked to the protection implemented). What we have retained in 20 years of research in the field is the importance of having a global approach for the protection, as much at the component level as for the electronic board.

General rules for a global ESD protection strategy

General rules

A first key point in having a global protection approach is the *consideration of ESD constraints starting with the specifications* of the system[1], by evaluating the level of robustness needed for the application being considered in relation to the environment in which it is to evolve, such as temperature considerations, for example. From these specifications, the ESD design window can be defined for each connection of the system with the outside.

A global protection approach also means having a *co-design approach* for the system for optimizing the protections as well as controlling the cost of their

---

1 Here a system is considered to be a component or an electronic board.

implementation. Moreover, the system environment can also affect the value of the final robustness. This approach allows consideration of:

– self-protection of some pins or connectors as long as this does not harm performance levels;

– taking advantage of the specificity of the circuit being protected:

- let us take the example of RF or high frequency circuits. Some of these circuits have no ESD protection as the use of active components for ESD protection leads to a degradation of the performance via their parasitic capacitance, and to problems of impedance mismatching. An example of an alternative design approach [KLE 00] involves relying on the passive components available in the technology (inductances, capacitances, transmission lines). This approach ensures a certain level of ESD robustness, while maintaining good performance for high frequencies,

- a similar approach can be used in low-noise analog circuits such as the LNA [THI 04];

– a choice and size of protection that fits the robustness level required (minimal silicon area) and the constraints of the performance (type of protection).

Rules for the protection strategy of an integrated circuit

During the design of the global ESD protection strategy of the IC, the ground rules are:

– each pin must have at least one ESD protection, with the exception of self-protected pins;

– there must be a discharge path between all of the couples of the pins and for different polarities, positive and negative.

Another aspect involves the sizing of the protection, as well as its placement and associated routing:

– the sizing of the protection is conditioned by the value of its *on-resistance* R_{ON}. The silicon footprint of the protection depends in part on the targeted ESD robustness, keeping in mind the impact of the operating temperature on the value of R_{ON};

– for Π protections, the *isolation resistance* must be big enough to limit the ESD current in the center of the circuit and its impact on circuit performance must also be taken into account during the simulations;

– the resistors in these protections are usually made of *polysilicon* to avoid the formation of any parasitic components. However, it is important to size them correctly, making sure that the *current capability* does not become the weak link of the protection;

– the same precaution must be applied to the *metallic rails* whose routing must be as short and simple as possible, so as to avoid the effects of high current densities and of localized thermal heating;

– the placement and number of *contacts and vias* must also be carried out carefully, checking their respective current capability;

– in digital technologies, the ESD pads are already available, and placement is therefore pre-imposed. In mixed and analog technologies, each new circuit requires a new approach for protection. Careful attention must be paid to the *placement of ESD protections* to avoid the formation of parasitic structures with certain blocks of the circuit;

– lastly, once the entire ESD protection strategy is available, it is important to simulate its impact on the function of the circuit (parasitic capacitances), as well as the reverse (accidental protection triggering by dV/dT or dI/dt).

A particularly important rule involves the *portability of ESD protection solutions*. This portability, which is highly sought after, is not as trivial as it seems. It is not so easy to transpose an ESD solution between different integrated circuits within the same technology, and even less between different technologies. The ESD expert is vital in this portability step.

In a given technology, the *cell library* also contains ESD protection elements. Before using an ESD protection structure, it is important to fully understand its operation, and to make sure that it is adapted for the pin being protected in terms of robustness, capacitance, leakage current, dynamics (response to a CDM type stress, for example), noise level, impedance, sensitivity to dV/dt or dI/dt, sensitivity to process variations or temperature, etc.

Rules for the protection strategy of an electronic board

In recent years, and in the face of the increasing complexity of the embedded applications, the robustness of systems to ESD events has become a priority. The amount of information that must be processed to carry out a predictive simulation of the robustness is a limitation. The parameters that make the reliability evaluation difficult are:

– Information not available due its proprietary nature. Each component manufacturer optimizes the protections so as to make the component more robust. The information that can be used in the simulation is therefore tied in with issues of intellectual property.

– The number of elements of the board system, which is always increasing. As we mentioned in the system approach presented in Chapter 3, interconnections and passive elements behave like non-linear elements, making their interactions hard to predict. Analysis of the discharge path in electronic boards requires a wide range of engineering techniques, thus resulting in a large number of skills to master.

– The lack of investigation techniques. When an issue of reliability due to an ESD stress is identified in the electronic boards, there is no investigation technique available that can help understand the causes of the failure, and therefore no means to come up with a solution.

These are the main challenges that we have tried to deal with in the research that we have carried out and presented in this work. By developing a complete methodology, from component to system, it has become possible to anticipate failures, whether material or functional. This methodology also involves a systemic approach where the component is considered as a black box. We believe behavioral modeling to be well-adapted as it allows intellectual property to be preserved for the protection strategies of circuit manufacturers, but it also deals with the issues involving the large complexity of system simulations.

Protections in integrated circuits play an important role in the prediction of the failure of a system caused by electrostatic discharge, whether this is material or functional. We first showed that the compact and behavioral models give roughly the same results during simulations, and this is true for both quasi-static and transient simulations. Behavioral models are based on quasi-static measurements obtained, thanks to a TLP bench. This modeling approach lets us carry out predictive simulations of failure with good levels of precision and reasonable calculation times, while ensuring good convergence.

A criticism of the approach can be made regarding the dynamic behavior of the protection during triggering, as it may introduce over-voltages that could potentially destroy the component being protected. In these models, the dynamic behavior of the protections around inflexion and snapback points is not always strictly reproduced. The voltage between the terminals of the component can then rise to well above the voltages extracted using TLP measurement. This argument is perfectly valid; however, in a system configuration, all of the parasitic elements linked to the interconnections and the package can slow down the rising edges of the ESD stress at the level of the chip. The addition of the parasitic elements of the IBIS partially solves this issue. Indeed, the parasitic capacitances of the inputs and outputs, as well as those located between the power supply rails impose their dynamics on the protections (the latter usually being quicker). These over-voltages still remain the cause of around 10% of the destruction observed. For the case when, despite all the measures taken, the triggering speed becomes non-negligible, we are working on the design of dynamic behavioral models [ESC 16a].

The integration of a large number of models, parameters and failure criteria are vital steps in the prediction of failures at the system level.

The following figure summarizes all of the elements to be considered in a system simulation, as well as the tools for creating a model for the detection of functional and material failures. The model of the integrated circuit must be added to the whole system in order to carry out the simulation properly. TLP measurements are vital for extracting a quasi-static model and validating the protection strategy of the circuit. The IBIS models provide information on the necessary parasitic elements of the inputs and inputs. Depending on the configuration, these can generate over-voltages and limit the circulation of transient currents. Beyond the data available in the datasheet, a HDL model (behavioral) of the circuit function is needed to properly carry out a functional simulation. Finally, the failure criteria that must be added to the models depend on the simulation required. A Wunsch & Bell model lets us predict the thermal destruction of a protection. For a functional analysis, other criteria can be added depending on the malfunction being detected. In this case, a probabilistic approach seems most appropriate, considering the random aspect of the ESD event in terms of the operating frequencies of digital systems [MON 13a, MON 11b].

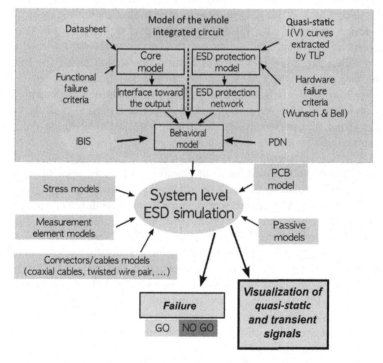

Figure 1. *Global ESD protection strategy. For a color version of this figure, see www.iste.co.uk/bafleur/esd.zip*

Conclusion

Thanks to the research carried out since the 1980s, the protection of integrated circuits from ESD is relatively well controlled. However, the appearance of a new technological generation every 18–24 months, with smaller feature size than the previous and related semiconductor volumes even smaller to dissipate the energy of these stresses, has pushed the components to their limits in terms of ESD robustness.

In order to deal with this issue, in 2007, the Industry Council on ESD Target Levels proposed to reduce the standard robustness of these components to 1 kV, arguing that the robustness of an electronic system does not only rely on the intrinsic robustness of the components that make it up.

This change marked a strong surge in interest regarding a global protection approach for electronic boards, which is vital in order to provide them with the required level of ESD robustness.

In this book, we have presented this approach, as well as the tools needed to put it into place. In this context, the "Holy Grail" of the system designer would be to have the design tools of an ESD module in their toolbox, allowing them to follow this co-design approach. This would require compact or behavioral models of the protections and of the application environment (stress waveforms), extraction tools for the possible parasitic structures and specific verification tools. The EDA Tool Working Group, from the ESDA, run by Michael G. Khazhinsky, has listed the requirements of ESD verification and the approaches already implemented in the design environment of several semiconductor companies [KHA 12]. At the same time, the group WG26 from ESDA, started in May 2014, works on the issue of component models for system level simulation.

This is the first book issued from our research team on the issue of electrostatic discharges in electronic systems. It is original both in its content and layout, compared to existing international works, since it covers both the component level, or the integrated circuit, and the system level, or electronic board. A significant section is dedicated to simulation, a crucial tool for research into optimizing protections for the system against ESD. Different simulation levels are tackled, depending on the desired granularity of the results. Some are closely linked to the physics of the phenomenon, which is very important for the optimization of the protection structures. Some levels are completely detached from the physics, but allow carrying out simulations on a much larger scale, and to do so without knowledge of the protection structure, which is often proprietary. Finally, based on twenty years of experience in ESD, this work presents several case studies, through which the reader can discover the concrete issues involving ESD, and perhaps begin to solve their own problems in the field.

The results from a number of PhD Theses carried out in our laboratory are summarized in this work. The manuscripts of these works are nearly all available in the open archives database of the CNRS (https://hal.archives-ouvertes.fr/). We would once again like to thank all of our doctoral students, most of whom have gone on to join various semiconductor companies or electronic manufacturers across the world (France, Netherlands, Switzerland, Austria, USA, etc.) as ESD experts, to our utmost satisfaction.

Lastly, we must stress that the issue of ESD is tightly linked to industrial applications. Thanks to the many projects carried out alongside our industry partners, we have acquired the equipment and experience necessary to become the national reference laboratory in the field.

Bibliography

[ABO 11] ABOUDA K., BESSE P., ROLLAND E., "Impact of ESD strategy on EMC performances: conducted emission and DPI immunity", *8th* IEEE *Workshop on Electromagnetic Compatibility of Integrated Circuits (EMC Compo)*, pp. 224–229, 2011.

[AGI 00] AGILENT TECHNOLOGIES, Time domain reflectometry theory, Application Note 1304-2, ref.5966-4855E, 17 pages, 2000.

[AIM 07] AIME J., ROUDET J., CLAVEL E. *et al.*, "Prediction and measurement of the magnetic near field of a static converter", *IEEE International Symposium on Industrial Electronics, 2007*, ISIE, pp. 2550–2555, 2007.

[ALA 08] ALAELDINE A., LACRAMPE N., BOYER A. *et al.*, "Comparison among emission and susceptibility reduction techniques for electromagnetic interference in digital integrated circuits", *Microelectronics Reliability*, vol. 39, no. 12, pp. 1728–1735, 2008.

[ANR 12] ANR, Projet ANR E-SAFE, "Système Electronique Automobile Robuste aux ESD", (2009–2012), available at: http://www.predit.prd.fr/predit4/projet.fo?cmd= creepdf&incDe=40498, 2012.

[ANS 99a] ANSI/ESD STM5.2-1999, Electrostatic Discharge Sensitivity Testing–Machine Model (MM) Component Level, 1999.

[ANS 99b] ANSI/ESD STM5.3.1, Charged Device Model (CDM)–Component Level, 1999.

[ANS 07] ANSI/ESD SP5.5.2-2007, Electrostatic Discharge Sensitivity Testing – Very Fast Transmission Line Pulse (VF-TLP) – Component Level, 2007.

[ANS 10] ANSI/ESDA/JEDEC JS-001, Electrostatic Discharge (ESD) sensitivity testing Human Body Model (HBM), 2010.

[ANS 14a] ANSI/ESDA/JEDEC JS-001-2014, Joint Standard for Electrostatic Discharge Sensitivity Testing – Human Body Model (HBM) – Component Level, August 2014.

[ANS 14b] ANSI/ESD STM5.5.1, Electrostatic Discharge Sensitivity Testing – Transmission Line Pulse (TLP) – Component Level, 2014.

[ANS 15a] ANSI/ESDA/JEDEC JS-002-2014, Joint Standard for Electrostatic Discharge Sensitivity Testing – Charged Device Model (CDM) – Device Level, April 2015.

[ANS 15b] ANSI/ESD SP14.5-2015, Near Field Immunity Scanning – Component/ Module/PCB Level, 2015.

[AOU 08] AOUINE O., LABARRE C., COSTA F., "Measurement and modeling of the magnetic near field radiated by a buck chopper", *IEEE Transactions on Electromagnetic Compatibility*, vol. 50, no. 2, pp. 445–449, 2008.

[ARB 12] ARBESS H., Structures MOS–IGBT sur technologie SOI en vue de l'amélioration des performances à haute température de composants de puissance et de protections ESD, PhD Thesis, Université Paul Sabatier, Toulouse III, France, 22 May 2012.

[ARB 14] ARBESS H., BAFLEUR M., TRÉMOUILLES D. *et al.*, "Combined MOS–IGBT–SCR structure for a compact high-robustness ESD power clamp in smart power SOI technology", *IEEE Transactions on Device and Materials Reliability*, vol. 14, no. 1, pp. 432–440, 2014.

[ARB 15] ARBESS H., BAFLEUR M., TRÉMOUILLES D. *et al.*, "Optimization of a MOS–IGBT–SCR ESD protection component in smart power SOI technology", *Microelectronics Reliability*, vol. 55, no. 9, pp. 1476–1480, 2015.

[ARO 84] ARORA V.K., "High-field electron mobility and temperature in bulk semiconductors", *Physical Review B Condensed Matter*, vol. 30, no. 12, pp. 7297–7298, 1984.

[BAF 10] BAFLEUR M., DAVENEL F., GUERVENO J.P. *et al.*, Guide d'analyse dans le cas de surcharge de type EOS et ESD, Guide ANADEF, September 2010.

[BAF 13] BAFLEUR M., CAIGNET F., NOLHIER N. *et al.*, "Tackling the challenges of system level ESD: from efficient ICs ESD protection to system level predictive modeling", *Taiwan ESD and Reliability Conference (TESDC)*, pp. 1–8, November 2013.

[BAK 90] BAKOGLU H.B., *Circuits, Interconnections, and Packaging for VLSI*, Addison-Wesley, 1990.

[BAL 96] BALIGA B.J., *Power Semiconductor Devices*, PWS Publishing Co, Boston, 1996.

[BAR 01] BARY L., BORGARINO M., PLANA R. *et al.*, "Transimpedance amplifier-based full low-frequency noise characterization setup for Si/SiGe HBTs", *IEEE Transactions on Electron Devices*, vol. 48, pp. 767–772, 2001.

[BAR 04] BARTH J., RICHNER J., HENRY L.G. *et al.*, "Real HBM and MM waveform parameters", *Journal of Electrostatics*, vol. 62, pp. 195–209, 2004.

[BEA 03] BEAUCHÊNE T., DEAN L., BEAUDOIN F. *et al.*, "Thermal laser stimulation and NB-OBIC techniques applied to ESD defect localization", *Microelectronics Reliability*, vol. 43, no. 3, pp. 439–444, 2003.

[BÈG 14a] BÈGES R., CAIGNET F., NOLHIER N. *et al.*, "Practical transient system-level ESD modeling – environment contribution", *EOS/ESD 2014 36th Annual Electrical Overstress/Electrostatic Discharge Symposium*, 2014.

[BÈG 14b] BÈGES R., CAIGNET F., BAFLEUR M. *et al.*, "Transient system-level ESD modeling of an automotive voltage regulator", *International ESD Workshop*, IEW, May 2014.

[BÈG 15] BÈGES R., CAIGNET F., BESSE P. *et al.*, "TLP-based human metal model stress generator and analysis method of ESD generators", *Electrical Overstress/Electrostatic Discharge Symposium (EOS/ESD 2015)*, Reno, September 2015.

[BER 99] BERNIER J.C., CROFT G.D., YOUNG W.R., "A process independent ESD design methodology", *Proceedings of the International Symposium on Circuits and Systems*, ISCAS'99, vol. 1, pp. 218–221, 1999.

[BER 01] BERTRAND G., "Conception et modélisation électrique de structures de protection contre les décharges électrostatiques en technologies BiCMOS et CMOS analogique", PhD Thesis, Institut National des Sciences Appliquées, Toulouse (France), 20 July 2001.

[BER 12] BERTONNAUD S., DUVVURY C., JAHANZEB A., "IEC system level ESD challenges and effective protection strategy for USB2 interface", *34rd Electrical Overstress/ Electrostatic Discharge Symposium (EOS/ESD)*, 2012.

[BES 02] BESSE P., ZECRI M., NOLHIER N. *et al.*, "Investigations for a self-protected LDMOS under ESD stress through geometry and design considerations for automotive applications", *Electrical Overstress/Electrostatic Discharge Symposium*, Charlotte, NC, pp. 348–353, 2002.

[BES 10] BESSE P., "ESD/EMC in an automotive environment", *Seminar Presented at IEW 2010*, Tutzing, Germany, May 2010.

[BES 11] BESSE P., LAFON F., MONNEREAU N. *et al.*, "ESD system level characterization and modeling methods applied to a LIN transceiver", *Proceeding of the Electrical Overstress/Electrostatic Discharge Symposium (EOS/ESD)*, Anaheim, CA, pp. 329–337, 11–16 September 2011.

[BOU 11] BOURGEAT J., Etude du thyristor en technologies CMOS avancées pour implémentation dans des stratégies locale et globale de protection contre les décharges électrostatiques, PhD Thesis, Paul Sabatier University, Toulouse III, France, 8 June 2011.

[BOY 07] BOYER A., Méthode de Prédiction de la Compatibilité Electromagnétique des Systèmes en Boîtier, PhD Thesis, Institut National des Sciences Appliquées de Toulouse, 2007.

[BRO 01] BROOKS R., "A simple model for a cable discharge event", *IEEE802.3 Cable Discharge Ad-hoc*, pp. 1–16, March 2001.

[CAI 10] CAIGNET F., MONNEREAU N., NOLHIER N., "Non-invasive system level ESD current measurement using magnetic field probe", *International Electrostatic Discharge Workshop 2010*, Tutzing, Germany, 10–13 May 2010.

[CAI 12] CAIGNET F., MONNEREAU N., NOLHIER N. *et al.*, "Behavioral ESD protection modeling to perform system level ESD efficient design", *2012 Asia-Pacific Symposium on Electromagnetic Compatibility (APEMC)*, pp. 401–404, 21–24 May 2012.

[CAI 13] CAIGNET F., NOLHIER N., WANG A. *et al.*, "20GHz on-chip measurement of ESD waveform for system level analysis", *Electrical Overstress/Electrostatic Discharge Symposium (EOS/ESD)*, pp. 1–9, 10–12 September 2013.

[CAI 15a] CAIGNET F., BEGES R., BESSE P. *et al.*, "Hierarchical modeling approach for system level ESD analysis: from hard to functional failure", *Asia-Pacific International Symposium on Electromagnetic Compatibility (APEMC)*, Taipei, May 2015.

[CAI 15b] CAIGNET F., NOLHIER N., BAFLEUR M., "Dynamic system level ESD current measurement using magnetic field probe", *Asia-Pacific International Symposium on Electromagnetic Compatibility (APEMC)*, Taipei, May 2015.

[CAN 06] CANIGGIA S., MARADEI F., "Circuit and numerical modeling of electrostatic discharge generators", *IEEE Transactions on Industry Applications*, vol. 42, no. 6, pp. 1350–1357, 2006.

[CAO 10] CAO Y., JOHNSSON D., ARNDT B. *et al.*, "A TLP-based human metal model ESD-generator for device qualification according to IEC 61000-4-2", *Asia-Pacific Symposium on Electromagnetic Compatibility (APEMC)*, pp. 471–474, 2010.

[CHA 90] CHARITAT G., Modélisation et réalisation de composants planar haute-tension. PhD Thesis, Paul Sabatier Toulouse University, 28 September 1990.

[CHA 13] CHARRUAU S., *Electromagnetism and Interconnections: Advanced Mathematical Tools for Computer-aided Simulation*, ISTE Ltd, London and John Wiley & Sons, New York, 2013.

[CHA 06] CHANG L., Cable discharge event, Application Note 1511, National Semiconductor, July 2006.

[CHE 07] CHEN S.-H., KER M.-D., "Optimization of PMOS-triggered SCR devices for on-chip ESD protection in a 0.18-µm CMOS technology", *IPFA 2007 14th International Symposium on the Physical and Failure Analysis of Integrated Circuits*, 2007.

[CHO 72] CHOO S.C., "Theory of a forward-biased diffused-junction P-L-N rectifier. I. Exact numerical solutions", *IEEE Transactions on Electron Devices*, vol. 19, no. 8, pp. 954–66, 1972.

[CHU 03] CHUI K.M., Simulation and measurement of ESD test for electronics devices, PhD Dissertation, 2003.

[CHU 04] CHUNDRU R., POMMERENKE D., WANG K. *et al.*, "Characterization of human metal ESD reference discharge event and correlation of generator parameters to failure levels-part I: reference event", *IEEE Transactions on Electromagnetic Compatibility*, vol. 46, no. 4, pp. 498–504, 2004.

[CLA 92] CLAYTON R.P., *Introduction to Electromagnetic Compatibility*, John Wiley & Sons Inc., Hoboken, 1992.

[COL 99] COLE E.I., TANGYUNYONG P., BENSON D.A. *et al.*, "TIVA and SEI developments for enhanced front and backside interconnection failure analysis", *Microelectronics Reliability*, vol. 39, no. 6, pp. 991–996, 1999.

[CUN 96] CUNY R.H.G., "SPICE and IBIS modeling kits the basis for signal integrity analyses", *International Symposium on Electromagnetic Compatibility*, pp. 204–208, 1996.

[DAS 96] DASCHER D.J., "Measuring parasitic capacitance and inductance using TDR", *Hewlett Packard Journal*, vol. 47, pp. 83–96, 1996.

[DEL 10] DELMAS A., GENDRON A., BAFLEUR M. *et al.*, "Transient voltage overshoots of high voltage ESD protections based on bipolar transistors in smart power technology", *Proceedings of the IEEE Bipolar/BiCMOS Circuits and Technology Meeting*, pp. 253–256, 2010.

[DEL 12] DELMAS A., Étude transitoire du déclenchement de protections haute tension contre les décharges électrostatiques, PhD Thesis, Paul Sabatier University, Toulouse III, France, 27 February 2012.

[DEM 06] DEMARTY S., Contribution à l'étude électromagnétique théorique et expérimentale des cartes de circuit imprimé, PhD dissertation, Limoges University, 2006.

[DI 07] DI SARRO J., CHATTY K., GAUTHIER R. *et al.*, "Evaluation of SCR-based ESD protection devices in 90 nm and 65 nm CMOS technologies", *IEEE International Reliability Physics Symposium Proceedings. 45th Annual*, 2007.

[DUT 75] DUTTON R.W., "Bipolar transistor modeling of avalanche generation for computer circuit simulation", *IEEE Transactions on Electron Devices*, vol. 22, no. 6, pp. 334–338, 1975.

[DUV 89] DUVVURY C., TAYLOR T., LINDGREN J. *et al.*, "Input protection design for overall chip reliability", *Proceedings of EOS/ESD Symposium*, pp. 190–197, 1989.

[EOS 16] EOS/ESD ASSOCIATION, "Electrostatic Discharge (ESD) Technology Road Map", EOS/ESD Association, available at: https://www.esda.org/assets/Uploads/docs/2016ESDATechnologyRoadmap.pdf, 2016.

[ESC 16a] ESCUDIÉ F., CAIGNET F., NOLHIER N. *et al.*, "From quasi-static to transient system level ESD simulation: extraction of turn-on elements", *Electrical Overstress/Electrostatic Discharge Symposium (EOS/ESD 2015)*, Anaheim, CA, September 2016.

[ESC 16b] ESCUDIÉ F., CAIGNET F., NOLHIER N. *et al.*, "Impact on non-linear capacitances on transient waveforms during system level ESD", *27th European Symposium on the Reliability of Electron Devices, Failure Physics and Analysis (ESREF)*, Berlin, Germany, September 2016.

[ESD 99] ESD ASSOCIATION, ESD STM5.3.1-1999: Standard Test Method for Electrostatic Discharge Sensitivity Testing – Charged Device Model (CDM) Component Level, 1999.

[ESD 01] ESD ASSOCIATION WG 5.1, ESD STM5.1-2001: Standard Test Method for Electrostatic Discharge Sensitivity Testing – Human Body Model (HBM) Component Level, 2001.

[ESD 09] ESD ASSOCIATION, Draft Standard ESD DSP5.6-2009. Human Metal Model (HMM), Component Level of ESD Association Working Group WG 5.6, 2009.

[ESM 01] ESMARK K., STADLER W., WENDEL M. *et al.*, "Advanced 2D/3D ESD device simulation – a powerful tool already used in a pre-Si phase", *Microelectronics Reliability*, vol. 41, no. 11, pp. 1761–70, 2001.

[ESM 03] ESMARK K., GOSSNER H., STADLER W., *Advanced Simulation Methods for ESD Protection Development*, Elsevier, Boston, 2003.

[ESS 04] ESSELY F., TREMOUILLES D., GUITARD N. *et al.*, "Study of the impact of multiple ESD stresses", *2nd Workshop EOS/ESD/EMI*, Toulouse, France, pp. 35–37, 2004.

[ESS 06] ESSELY F., DARRACQ F., POUGET V. *et al.*, "Application of various optical techniques for ESD defect localization", *Microelectronics Reliability*, vol. 46, no. 9, pp. 1563–1568, 2006.

[ETH 15] ETHERTON M., RUTH S., MILLER J.W. *et al.*, "A new full-chip verification methodology to prevent CDM oxide failures", *37th Electrical Overstress/Electrostatic Discharge Symposium*, September 2015.

[FLE 57] FLETCHER N.H., *The High Current Limit for Semiconductor Junction Devices*, Institute of Radio Engineers, pp. 862–72, 1957.

[FLE 88] FLETCHER M., ABEL A., WAHID P.F. *et al.*, "Modeling of crosstalk in coupled microstrip lines", *Southeastcon'88, IEEE Conference Proceedings*, pp. 506–510, 1988.

[FUK 04] FUKUI S., NAOI T., TOYAMA N., "ESD current measurement using the near mangetic field", *Journal of SAE Transactions*, vol. 113, no. 7, pp. 250–256, 2004.

[GAL 02] GALY P., BERLAND V., FOUCHER B. *et al.*, "Experimental and 3D simulation correlation of a gg-nMOS transistor under high current pulse", *Microelectronics Reliability*, vol. 42, nos. 9–11, pp. 1299–1302, 2002.

[GAL 12] GALY P., ENTRINGER C., DRAY A., inventors; Stmicroelectronics SA, assignee. Circuit for protecting an integrated circuit against electrostatic discharges in CMOS technology, US patent 8,164,871., 24 April 2012.

[GAL 13] GALY P., ENTRINGER C., JIMENEZ J., inventors; Stmicroelectronics SA, assignee. Structure for protecting an integrated circuit against electrostatic discharges, US patent 8,610,216., 17 December 2013.

[GAO 07] GAO Y., GUITARD N., SALAMERO C. *et al.*, "Identification of the physical signatures of CDM induced latent defects into a DC–DC converter using low frequency noise measurements", *Microelectronics Reliability*, vol. 47, pp. 1466–1471, 2007.

[GAO 09] GAO Y., Stratégies de modélisation et protection vis-à-vis des décharges électrostatiques (ESD) adaptées aux exigences de la norme du composant charge (CDM), PhD Thesis, Institut National Polytechnique, Toulouse, 13 February 2009.

[GEN 07a] GENDRON A., Structures de protection innovantes contre les décharges électrostatiques dédiées aux entrées/sorties hautes tensions de technologies SmartPower, PhD Thesis, University Paul Sabatier, Toulouse, France, 29 March 2007.

[GEN 07b] GENDRON A., RENAUD P., BESSE P., Semiconductor device structure and integrated circuit therefor, Patent no. WO2007104342, 20 September 2007.

[GEV 15] GEVINTI E., CERATI L., DI BICCARI L. *et al.*, "Schematic-level and layout-level ESD EDA check methodology applied to smart power IC's – initialization and implementation", *37th Electrical Overstress/Electrostatic Discharge Symposium, 2015*, EOS/ESD, September 2015.

[GHA 85] GHARBI M., La tenue en tension et le calibre en courant du transistor MOS vertical dans la gamme des moyennes tensions (300 à 1000 volts), PhD Thesis, Paul Sabatier University, Toulouse, 1985.

[GIE 98] GIESER H., HAUNSCHILD M., "Very fast transmission line pulsing of integrated structures and the charged device model", *IEEE Transactions on Components, Packaging and Manufacturing Technology*, vol. 21, no. 4, pp. 278–285, 1998.

[GIR 10] GIRALDO S. SALAMERO C., CAIGNET F., "Impact of the power supply on the ESD system level robustness", *Electrical Overstress/Electrostatic Discharge Symposium (EOS/ESD), 2010 32nd*, pp. 1–8, 2010.

[GIR 13] GIRALDO TORRES S., Etude de la robustesse d'amplificateurs embarqués dans des applications portables soumis à des décharges électrostatiques (ESD) au niveau système, PhD Thesis, University of Toulouse, 2013.

[GOS 02] GOSSNER H., Integrated semiconductor circuit with protective structure for protection against electrostatic discharge, US Patent 6441437 B1, 27 August 2002.

[GRE 91] GREASON W.D., "Electrostatic discharge: a charged driven phenomenon", *Electrical Overstress/Electrostatic Discharge Symposium (EOS/ESD)*, vol. 28, pp. 199–218, 1991.

[GRE 02] GREASON W.D., "Generalized model of electrostatic discharge (ESD) for bodies in approach: analyses of multiple discharges and speed of approach", *Journal of Electrostatics*, vol. 54, pp. 23–37, 2002.

[GRU 04] GRUND E., GAUTHIER R., "VF-TLP systems using TDT and TDRT for kelvin wafer measurements and package level testing", *EOS/ESD 2004 26th Annual Electrical Overstress/Electrostatic Discharge Symposium*, 2004.

[GUI 04a] GUITARD N., TREMOUILLES D., ALVES S. *et al.*, "ESD induced latent defects in CMOS ICs and reliability impact", *EOS/ESD 2004 26th Annual Electrical Overstress/Electrostatic Discharge Symposium*, pp. 174–181, 2004.

[GUI 04b] GUITARD N., TREMOUILLES D., BAFLEUR M. *et al.*, "Low frequency noise measurements for ESD latent defect detection in high reliability applications", *Microelectronics Reliability*, vol. 44, nos. 9–11, pp. 1781–1786, September–November 2004.

[GUI 05] GUITARD N., TREMOUILLES D., ESSELY F. *et al.*, "Different failure signatures of multiple TLP and HBM stresses in an ESD robust protection structure", *Microelectronics Reliability*, vol. 45, nos. 9–10, pp. 1415–1420, September–November 2005.

[GUI 06] GUITARD N., Caractérisation de défauts latents dans les circuits intégrés soumis à des décharges électrostatiques, PhD Thesis, Paul Sabatier University, Toulouse, 26 October 2006.

[GUM 70] GUMMEL H., POON H., "An integral charge control model of bipolar transistors", *Bell System Technical Journal*, vol. 49, no. 5, pp. 827–852, 1970.

[HAL 09] HALL S.H., HECK H.L., *Advanced Signal Integrity for High-Speed Digital Designs*, Wiley, 2009.

[HAM 05] HAMON J.C., Méthodes et outils de la conception amont pour les systèmes et les microsystèmes, PhD Thesis, Institut National Polytechnique de Toulouse, France, 2005.

[HAN 07] HAN L., KOO J., POMMERENKE D. *et al.*, "Experimental investigation of the ESD sensitivity of an 8-bit microcontroller", *IEEE International Symposium on Electromagnetic Compatibility*, pp. 1–6, 2007.

[HON 93] HONG S., KIM J., YOO K. *et al.*, "Two-dimensional electrothermal simulations and design of electrostatic discharge protection circuit", *EOS/ESD Symposium*, Orlando, FL, pp. 157–163, 1993.

[HON 07] HONDA M., "Measurement of ESD-gun radiated fields", *29th Electrical Overstress/Electrostatic Discharge EOS/ESD Symposium*, 2007.

[HUA 10] HUANG W., LIU D., XIAO J. *et al.*, "Probe characterization and data process for current reconstruction by near field scanning method", *Symposium on Electromagnetic Compatibility (APEMC), 2010 Asia-Pacific*, pp. 467–470, 2010.

[HYA 02] HYATT H., "ESD: standards, threats and system hardness fallacies", *EOS/ESD 2002 24th Annual Electrical Overstress/Electrostatic Discharge Symposium*, pp. 178–185, 2002.

[IBI 95] IBIS (Input Output Buffer Information Specification), ed. 1, which has been formally ratified as ANSI/EIA-656 on December 13, 1995, www.eigroup.org/IBIS, 1995.

[IBI 13] IBIS (Input Output Buffer Information Specification), ed. 6, which has been formally ratified as ANSI/EIA-656 on September 20, 2013, www.eigroup.org/IBIS, 2013.

[IEC 05] IEC, Electromagnetic Compatibility (EMC), Part 4-20: Testing and Measurement Techniques, Emission and Immunity Testing in Transverse Electromagnetic (TEM) Waveguides, Basic EMC publication, IEC 61000-4-20, 2005.

[IEC 06] IEC, Electromagnetic Compatibility (EMC), Integrated Circuits – Measurement of Electromagnetic Immunity, 150 kHz to 1 GHz – Part 1: General Conditions and Definitions, IEC 62132, January 2006.

[IEC 07a] IEC, Electromagnetic Compatibility (EMC), Integrated Circuits, Measurement of Electromagnetic Emissions, 150 kHz to 1 GHz – Part 4: Measurement of Conducted Emissions – 1Ω/150Ω Direct Coupling Method, IEC61967-4, 2007.

[IEC 07b] IEC, Electromagnetic Compatibility (EMC), Integrated Circuits, Measurement of Electromagnetic Immunity – 150 kHz to 1 GHz – Part 3: Bulk Current Injection (BCI) Method, IEC62132-3, 2007.

[IEC 08] IEC, Electromagnetic Compatibility (EMC) – Part 4-2: Testing and Measurement Techniques – Electrostatic Discharge Immunity Test, IEC-61000-4-2, Edition 2.0, 1 December 2008.

[IEC 10] IEC, Compatibilité électromagnétique (CEM). Modèles de circuits intégrés pour la simulation du comportement lors de perturbations électromagnétiques. Modélisation des émissions conduites (ICEM-CE), IEC 62433-2, IEC Standard, February 2010.

[IEC 14] IEC, Electromagnetic Compatibility (EMC), Integrated Circuits – Measurement of Electromagnetic Emissions – Part 3: Measurement of Radiated Emissions – Surface Scan Method, 61967-3, ed. 2, 2014.

[IEE 08] IEEE, Standard IEEE 1076.1: VHDL Analog and Mixed-Signal, IEEE, 2008.

[IND 10] INDUSTRY COUNCIL ON ESD TARGET LEVELS ESDA WORKING GROUP, White Paper 3: System Level ESD Part I: Common Misconceptions and Recommended Basic Approaches, December 2010.

[IND 13] INDUSTRY COUNCIL ON ESD TARGET LEVELS, White Paper 1: A Case for Lowering Component Level HBM/MM ESD Specifications and Requirements, available at: www.esdindustrycouncil.org, 25 June 2013.

[ISO 08] ISO, Road Vehicles – Test Methods for Electrical Disturbances from Electrostatic Discharge, ISO 10605, 2008-07-15, 2008.

[ITR 05] ITR, International Technology Roadmap for Semiconductors, Edition 2005, ITR, available at: www.itrs.net, 2005.

[IWA 93] IWATA H., AKAO Y., "Characteristics of e-field near indirect ESD events", *IEEE International Symposium on Electromagnetic Compatibility*, pp. 26–27, 1993.

[JAN 93] JANG S.L., "On the common-emitter breakdown voltage of bipolar junction transistors", *Solid State Electronics*, vol. 36, no. 2, pp. 213–216, 1993.

[JIA 11] JIANG R.H.-C., TSENG T.-K., CHEN C.-H. *et al.*, "Design of on-chip transient voltage suppressor in a silicon-based transceiver IC to meet IEC system-level ESD specification", *IEEE International Conference on IC Design & Technology (ICICDT)*, pp. 1–4, 2011.

[JIU 01] JIUSHENG H., QIBIN D., FANG LIU L. *et al.*, "Electromagnetic field generated by transient electrostatic discharges (ESD) from person charged with low electrostatic voltage", *23th Electrical Overstress/Electrostatic Discharge Symposium, EOS/ESD*, pp. 413–416, 2001.

[KAS 98] KASH J.A., TSANG J.C., RIZZOLO R.F. *et al.*, "Backside optical emission diagnostics for excess IDDQ", *IEEE Journal of Solid-State Circuits*, vol. 33, no. 3, pp. 508–511, 1998.

[KAS 99] KASH J.A., TSANG J.C., KNEBEL D.R. *et al.*, "Non-Invasive Backside Failure Analysis of Integrated Circuits by Time-Dependent Light Emission: Picosecond Imaging Circuit Analysis", *Microelectronic Failure Analysis*, pp. 505–510, 1999.

[KER 99a] KER M.-D., CHANG H.H., "How to safely apply the LVTSCR for CMOS whole-chip ESD protection without being accidentally triggered on", *Journal of Electrostatics*, vol. 47, no. 4, pp. 215–248, 1999.

[KER 99b] KER M.-D., CHANG H.-H., CHEN T.-Y., "ESD buses for whole-chip ESD protection", *Proceedings of the 1999 IEEE International Symposium on Circuits and Systems*, IEEE, pp. 545–548, 1999.

[KHA 12] KHAZHINSKY M.G., CAO S., GOSSNER H. *et al.*, "Electronic design automation (EDA) solutions for ESD-robust design and verification", *Proceedings of the IEEE 2012 Custom Integrated Circuits Conference*, pp. 1–8, September 2012.

[KHU 86] KHURANA N., CHIANG C.L., "Analysis of product hot electron problems by gated emission microscopy", *IRPS Proceedings, International Reliability Physics Symposium*, pp. 189–194, 1986.

[KLA 92] KLAASSEN D.B.M., SLOTBOOM J.W., DE GRAAFF H.C., "Unified apparent bandgap narrowing in n-and p-type silicon", *Solid-State Electronics*, vol. 35, no. 2, pp.125–129, 1992.

[KLE 00] KLEVELAND B., MALONEY T.J., MORGAN I. *et al.*, "Distributed ESD protection for high-speed integrated circuits", *IEEE Electron Device Letters*, vol. 21, no. 8, pp. 390–392, 2000.

[KOO 08] KOO J., CAI Q., WANG K. *et al.*, "Correlation between EUT failure levels and ESD generator parameters", *IEEE Transactions on Electromagnetic Compatibility*, vol. 50, no. 4, pp. 794–801, 2008.

[KOY 95] KOYAMA T., MASHIKO Y., SEKINE M. *et al.*, "New non-bias optical beam induced current (NB-OBIC) technique for evaluation of Al interconnects", *Proceedings of 1995 IEEE International Reliability Physics Symposium*, 1995.

[LAC 91] LACKNER T., "Avalanche multiplication in semiconductors: a modification of Chynoweth's law", *Solid State Electronics*, vol. 34, no. 1, pp. 33–42, 1991.

[LAC 06] LACRAMPE N., BOYER A., NOLHIER N. *et al.*, "Original methodology for integrated circuit ESD immunity combining VF-TLP and near field scan testing", *3rd EOS/ESD/EMI Workshop*, Toulouse, France, 18–19 May 2006.

[LAC 07a] LACRAMPE N., ALAELDINE A, CAIGNET F. *et al.*, "Investigation on ESD transient immunity of integrated circuits", *IEEE International Symposium on Electromagnetic Compatibility*, Honolulu, HI, 8–13 July 2007.

[LAC 07b] LACRAMPE N., CAIGNET F., BAFLEUR M. *et al.*, "Characterization and modeling methodology for IC's ESD susceptibility at system level using VF-TLP tester", *29th Electrical Overstress/Electrostatic Discharge Symposium, 2007*, pp. 5B.1-1, 5B.1-7, 16–21 September 2007.

[LAC 08] LACRAMPE N., Méthodologie de modélisation et de caractérisation de l'immunité des cartes électroniques vis-à-vis des décharges électrostatiques (ESD), PhD Dissertation, INSA, Toulouse, 2008.

[LAF 11] LAFON F., Développement de techniques et de méthodologies pour la prise en compte des contraintes CEM dans la conception d'équipements du domaine automobile. Etude de l'immunité, du composant à l'équipement, PhD Dissertation, INSA Rennes, 2011.

[LAI 06] LAI T.X., KER M.D., "Method to evaluate cable discharge event (CDE) reliability of integrated circuits in CMOS technology", *International Symposium on Quality Electronic Design (ISQED '06)*, pp. 597–602, 2006.

[LAU 91] LAURIN J.J., ZAKY S.G., BALMAIN K.G., "EMI-induced failures in crystal oscillators", *IEEE Transactions on Electromagnetic Compatibility*, vol. 33, pp. 334–342, 1991.

[LEE 10] LEE J.-H., SHIH J.R., KUAN H.P. *et al.*, "The influence of decoupling capacitor on the discharge behavior of fully silcided power-clamped device under HBM ESD event", *17th IEEE International Symposium on the Physical and Failure Analysis of Integrated Circuits (IPFA)*, 2010.

[LES 15] LESCOT J., DEHAN P., BOUJARRA W. *et al.*, "A comprehensive ESD verification flow at transistor level for large SoC designs", *37th Electrical Overstress/Electrostatic Discharge Symposium,* September 2015.

[LET 69] LETURCQ P., Comportement électrique et thermique des transistors bipolaires aux forts niveaux de tension et de courant : application au phénomène de second claquage. PhD Thesis, University of Toulouse, 10 October 1969.

[LIP 10] LIPPER C., "Future challenges of ESD protection – an automotive OEM's point of view", *Seminar Presented at International ESD Workshop (IEW) 2010*, Tutzing, Germany, 2010.

[LOM 88] LOMBARDI C., MANZINI S., SAPORITO A. *et al.*, "A physically based mobility model for numerical simulation of nonplanar devices", *IEEE Transactions on Computer Aided Design of Integrated Circuits and Systems*, vol. 7, no. 11, pp. 1164–1171, 1988.

[MAL 85] MALONEY T.J., KHURANA N., "Transmission line pulsing techniques for circuit modeling of ESD phenomena.", *7th Electrical Overstress/Electrostatic Discharge Symposium (EOS/ESD)*, vol. 7, pp. 49–54, September 1985.

[MAR 09] MARUM S., DUVVURY C., PARK J. *et al.*, "Protecting circuits from the transient voltage suppressor's residual pulse during IEC 61000-4-2 stress", *31st Electrical Overstress/Electrostatic Discharge Symposium (EOS/ESD)*, pp. 1–10, 2009.

[MAS 83] MASETTI G., SEVERI M., SOLMI S., "Modeling of carrier mobility against carrier concentration in arsenic-, phosphorus-, and boron-doped silicon", *IEEE Transactions on Electron Devices*, vol. 30, no. 7, pp. 764–769, 1983.

[MCA 96] MCANDREW C.C., SEITCHIK J.A., BOWERS D.F. *et al.*, "VBIC95, the vertical bipolar inter-company model", *IEEE Journal of Solid-State Circuits*, vol. 31, no. 10, pp. 1476–1483, 1996.

[MCM 00] MCMANUS M.K., KASH J.A., STEEN S.E. *et al.*, "PICA: backside failure analysis of CMOS circuits using picosecond imaging circuit analysis", *Microelectronic Reliability*, vol. 40, pp. 1353–1358, 2000.

[MER 93] MERRILL R., ISSAQ E., "ESD design methodology", *The Electrical Overstress/Electrostatic Discharge Symposium*, pp. 233–237, 1993.

[MER 12] MERTENS R., ROSENBAUM E., KUNZ H. *et al.*, "A flexible simulation model for system level ESD stresses with applications to ESD design and troubleshooting", *34th Electrical Overstress/Electrostatic Discharge Symposium (EOS/ESD)*, 2012.

[MIL 57] MILLER S., "Ionization rates for holes and electrons in silicon", *Physical Review*, vol. 105, no. 4, pp. 1246–1249, February 1957.

[MIL 89] MIL-STD-883^E, Method 3015.7, Electrostatic Discharge Sensitivity Classification, Department of Defence, Test Method Standard, Microcircuits, 1989.

[MON 10] MONNEREAU N., CAIGNET F., TREMOUILLES D. *et al.*, "Building-up of system level ESD modeling: impact of a decoupling capacitance on ESD propagation", *32nd Electrical Overstress/Electrostatic Discharge Symposium (EOS/ESD)*, 2010.

[MON 11a] MONNEREAU N., Développement d'une methodologie de caractérisation et de modélisation de l'impact des décharges électrostatiques sur les systèmes électroniques, PhD Dissertation, University of Toulouse, 2011.

[MON 11b] MONNEREAU N., CAIGNET F., NOLHIER N. *et al.*, "Investigating the probability of susceptibility failure within ESD system level consideration", *Electrical Overstress/Electrostatic Discharge Symposium (EOS/ESD), 2011*, pp. 1–6, 11–16 September 2011.

[MON 11c] MONNEREAU N., CAIGNET F., NOLHIER N. *et al.*, "Behavioral-modeling methodology to predict electrostatic-discharge susceptibility failures at system level: an IBIS improvement", *EMC Europe 2011*, York, pp. 457–463, 26–30 September 2011.

[MON 12] MONNEREAU N., CAIGNET F., NOLHIER N. *et al.*, "Investigation Of Modeling System ESD Failure And Probability Using IBIS ESD models", *IEEE Transactions on Device and Materials Reliability*, vol. 12, no. 4, pp. 599–606, December 2012.

[MON 13a] MONNEREAU N., CAIGNET F., TREMOUILLES D. *et al.*, "A system-level electrostatic-discharge-protection modeling methodology for time-domain analysis", *IEEE Transactions on EMC*, vol. 55, no. 1, pp. 45–57, February 2013.

[MON 13b] MONNEREAU N., CAIGNET F., TREMOUILLES D. *et al.*, "Building-up of system level ESD modeling: impact of a decoupling capacitance on ESD propagation", *Microelectronics Reliability*, vol. 53, no. 2, pp. 221–228, February 2013.

[MOO 65] MOORE G.E., "Cramming more components onto integrated circuits", *Electronics*, pp. 114–117, 19 April 1965.

[MUH 09] MUHONEN K., PEACHEY N., TESTIN A., "Human metal model (HMM) testing, challenges to using ESD guns", *EOS/ESD Symposium, 2009 31st*, pp. 1–9, 2009.

[NIK 98] NIKAWA K., SHOJI I., "Detection and characterization of failures and defects in LSI chips by optical beam induced resistance changes (OBIRCH)", *Conference Series-Institute of Physics*, vol. 160, pp. 37–46, 1998.

[NOL 94] NOLHIER N., STEFANOV E., CHARITAT G. *et al.*, "Power 2D: dedicated tool for two-dimensional simulation of off-state power structures", *International Journal for Computation and Mathematics Electrical and Electronic Engineering*, vol. 13, no. 4, pp. 771–783, 1994.

[NOL 02] NOLHIER N., BAFLEUR M., BESSE P. *et al.*, "Couplage physique/circuits pour la simulation "2D mixed mode" d'un événement ESD", *3ème Colloque Sur Le Traitement Analogique De l'Information Du Signal Et Ses Applications*, Paris, France, p. 5, 2002.

[OH 02] OH K.-H., DUVVURY C., BANERJEE K. *et al.*, "Investigation of gate to contact spacing effect on ESD robustness of salicided deep submicron single finger NMOS transistors", *Reliability Physics Symposium Proceedings, 2002, 40th Annual*, pp. 148–155, 2002.

[OKU 75] OKUTO Y., CROWELL C.R., "Threshold energy effects on avalanche breakdown voltage in semiconductor junction", *Solid State Electronics*, vol. 18, pp. 161–68, 1975.

[OLN 03] OLNEY A., GIFFORD B., GURAVAGE J. *et al.*, "Real-world charged board model (CBM) failures", *EOS/ESD 2003 25th Electrical Overstress/Electrostatic Discharge Symposium*, 2003.

[ORR 13] ORR B., MAHESHWARI P., POMMERENKE D. *et al.*, "A systematic method for determining soft-failure robustness of a subsystem", *35th Electrical Overstress/Electrostatic Discharge Symposium (EOS/ESD)*, 2013.

[OVE 70] OVERSTRAETEN R.V., MAN H.D., "Measurement of the ionization rates in diffused Silicon p-n junctions", *Solid State Electronics*, vol. 13, pp. 583–608, 1970.

[PER 96] PÉREZ J., CARLES R., FLECKINGER R., *Electromagnétisme, fondements et applications: avec 300 exercices et problèmes résolus*, Masson, 1996.

[PIN 84] PINTO M., RAFFERTY C., YEAGER H. *et al.*, Pisces-Ii - Poisson and Continuity Equation Solver, Report, Stanford Electronics Laboratory, 1984.

[PIS 05] PISCHL N., "ESD transfer through Ethernet magnetics", *International Symposium on Electromagnetic Compatibility*, vol. 2, pp. 356–363, 2005.

[POG 03] POGANY D., BYCHIKHIN S., GORNIK E. *et al.*, "Moving current filaments in ESD protection devices and their relation to electrical characteristics", *IEEE International Reliability Physics Symposium*, Dallas, TX, pp. 241–248, 2003.

[POM 95] POMMERENKE D., AIDAM M., "To what extent do contact-mode and indirect ESD test methods reproduce reality?", *Electrical Overstress/Electrostatic Discharge Symposium Proceedings*, pp. 101–109, 1995.

[REM 09] REMMACH M., Analyse de défaillance des circuits intégrés par émission de lumière dynamique: développement et optimisation d'un système expérimental, PhD Thesis, University of Bordeaux 1, 3 September 2009.

[REY 86] REYNES J.M., Relations entre performances et paramètres structuraux des transistors, Application à la conception des composants, PhD Thesis, Institut National des Sciences Appliquées, 25 March 1986.

[RIV 04] RIVENC J., VAZQUEZ-GARCIA J., MATOSSIAN P. et al., "An overview of the technical policy developed by Renault to manage ESD risks in airbags", Industry Applications Conference, 2004, 39th IAS Annual Meeting, vol. 2, pp. 1294–1301, 3–7 October 2004.

[RUS 98] RUSS C., BOCK K., RASRAS M. et al., "Non-uniform triggering of gg-NMOSt investigated by combined emission microscopy and transmission line pulsing", EOS/ESD 1998 15th Electrical Overstress/Electrostatic Discharge Symposium, pp. 177–186, 1998.

[RUS 99] RUSS C., ESD Protection Devices for CMOS Technologies: Processing Impact, Modeling and Testing Issues, ISBN 3-8265-6664-5, München, Techn. Univ., Diss., Shaker Verlag, 1999.

[SAE 12] SAE, Vehicle Architecture For Data Communications Standards – LIN Conformance Test Specification for LIN 2.0, Standard J2602-2, Version 2.0, November 2012.

[SAL 05a] SALAMERO C., NOLHIER N., BAFLEUR M. et al., "Accurate prediction of the ESD robustness of semiconductor devices through physical simulation", IEEE International Reliability Physics Symposium, San Jose, CA, pp. 106–111, 2005.

[SAL 05b] SALAMERO C., Prédiction de la robustesse des structures de protections ESD, PhD Thesis, Paul Sabatier University, Toulouse, 2005.

[SCH 12] SCHOLZ M., VANDERSTEEN G., SHIBKOV A. et al., "HMM single site testing: Can we reproduce component failure level with the HMM document?", 34rd Electrical Overstress/Electrostatic Discharge Symposium (EOS/ESD 2012), September 2012.

[SCH 16] SCHOLZ M., ASHTON R., SMEDES T. et al., "HMM single site testing: Can we reproduce component failure level with the HMM document?", Electrical Overstress/Electrostatic Discharge Symposium (EOS/ESD 2015), Anaheim, CA, September 2016.

[SEM 08] SEMENOV O., SOMOV S., "ESD protection design for I/O libraries in advanced CMOS technologies", Journal of Solid-State Electronics, vol. 52, no. 8, pp. 1127–1139, August 2008.

[SEN 16] SENTAURUS TCAD, "Release L-2016.03", available at: www.synopsys.com, April 2016.

[SIL 16] SILVACO TCAD, "Baseline Release", available at: www.silvaco.com, April 2016.

[SLO 77] SLOTBOOM J.W., DE GRAAF H.C., "Bandgap narrowing in silicon bipolar transistors", *IEEE Transactions on Electron Devices*, vol. 24, no. 8, pp. 1123–1125, 1977.

[SME 08] SMEDES T., CHRISTOFOROU Y., "On the relevance of IC ESD performance to product quality", *EOS/ESD 30th Electrical Overstress/Electrostatic Discharge Symposium*, 2008.

[SMI 02] SMITH C., "Cable effect part 1: cable discharge events", *Technical Tidbit*, High Frequency Measurements Web Page, January 2002.

[SMO 99] SMOLYANSKY D., COREY S., "PCB interconnect characterization from TDR measurements", *Electronic Engineering*, vol. 71, pp. 63–64, 1999.

[SPÉ 06] SPÉCIFICATION PRODUIT TEKTRONIC, AC Current Probes, CT1, CT2, CT6 Data Sheet, 2006.

[STA 98] STADLER W., GUGGENMOS X., EGGER P. *et al.*, "Does the ESD-failure current obtained by transmission-line pulsing always correlate to human body model tests?", *Microelectronics Reliability*, vol. 38, no. 11, pp.1773–1780, 1998.

[STA 07] STADLER W., "State-of-the-art in ESD standards", *International ESD Workshop*, May 2007.

[STA 09] STADLER W., BRODBECK T., GÄRTNER R. *et al.*, "Do ESD fails in systems correlate with IC ESD robustness?", *Microelectronics Reliability*, vol. 49, no. 9, pp. 1079–1085, 2009.

[STA 15] STADLER W., NIEMESHEIM J., GUERSES H. *et al.*, "Practical HBM testing with statistical pin combinations", *EOS/ESD 2015 37th Electrical Overstress/Electrostatic Discharge Symposium*, 2015.

[STO 03] STOCKINGER M., MILLER J.W., KHAZHINSKY M.G. *et al.*, "Boosted and distributed rail clamp networks for ESD protection in advanced CMOS technologies", *Electrical Overstress/Electrostatic Discharge Symposium*, pp. 3–21, 2003.

[STO 04] STOCKINGER M., MILLER J.W., "Advanced ESD rail clamp network design for high voltage CMOS applications", *Electrical Overstress/Electrostatic Discharge Symposium Proceedings*, vol. 26, p. 280, 2004.

[STO 05] STOCKINGER M., MILLER J.W., KHAZHINSKY M.G. *et al.*, "Advanced rail clamp networks for ESD protection", *Microelectronics Reliability*, vol. 45, no. 2, pp. 211–222, 2005.

[STR 00] STRICKER A., Technology computer aided design of ESD protection devices, Thesis, EHT, Zurich, 2000.

[SZE 81] SZE S.M., *Physics of Semiconductor Devices*, John Wiley, Chichester, 1981.

[TAS 70] TASCA D.M., "Pulse power failure modes in semiconductors", *IEEE Transactions on Nuclear Science*, vol. 17, no. 6, pp. 346–372, December 1970.

[THI 04] THIJS S., NATARAJAN M.I., LINTEN D. *et al.*, ESD protection for a 5.5 GHz LNA in 90 nm RF CMOS – implementation concepts, constraints and solutions", *Electrical Overstress/Electrostatic Discharge Symposium*, pp. 1–10, September 2004.

[THI 10] THIJS S., "System to component level correlation factor", *2010 IEW Workshop*, Tutzing, Germany, 2010.

[TOR 02] TORRES C.A., MILLER J.W., STOCKINGER M. *et al.*, "Modular, portable, and easily simulated ESD protection networks for advanced CMOS technologies", *Microelectronics Reliability*, vol. 42, no. 6, pp. 873–885, 2002.

[TRA 84] TRANDUC H., ROSSEL P., SANCHEZ J.-L., "Premier et second claquage dans les transistors MOS", *Revue de Physique Appliquée*, vol. 19, no. 10, pp. 859–878, 1984.

[TRE 02] TREMOUILLES D., BERTRAND G., BAFLEUR M. *et al.*, "Design guidelines to achieve a very high ESD robustness in a self-biased NPN", *EOS/ESD Symposium*, Charlotte, NC, pp. 281–288, 2002.

[TRE 04a] TREMOUILLES D., Optimisation et modélisation de protections intégrées contre les décharges électrostatiques, par l'analyse de la physique mise en jeu, PhD Thesis, Institut National des Sciences Appliquées de Toulouse, France, 14 May 2004.

[TRE 04b] TREMOUILLES D., BAFLEUR M., BERTRAND G. *et al.*, "Latch-up ring design guidelines to improve electrostatic discharge (ESD) protection scheme efficiency", *IEEE Journal of Solid-State Circuits*, vol. 39, no. 10, pp. 1778–1782, Octobre 2004.

[TRE 10] TREMOUILLES D., MONNEREAU N., CAIGNET F. *et al.*, "Simple ICs-internal-protection models for system level ESD simulation", *International Electrostatic Discharge Workshop (IEW 2010)*, Tutzing, Allemagne, 10–13 May 2010.

[VAL 99] VALDINOCI M., VENTURA D., VECCHI M.C. *et al.*, "Impact ionization in silicon at large operating temperature", *International Conference on Simulation of Semiconductor Process and Devices*, Kyoto, Japan, pp. 27–30, 1999.

[VAS 03] VASHCHENKO V.A., CONCANNON A., TER BEEK M. *et al.*, "Quasi-3D simulation approach for comparative evaluation of triggering ESD protection structures", *Microelectronics Reliability*, vol. 43, no. 3, pp. 427–437, 2003.

[VER 14] VERILOG, "AMS Language Reference Manual v2.4", available at: http://accellera.org/, 2014.

[VIN 98] VINSON J.E., LIOU J.J., "Electrostatic discharge in semiconductor devices: an overview", *Revue de physique appliquée*, vol. 86, no. 2, pp. 399–418, February 1998.

[VLA 94] VLADIMIRESCU A., *The SPICE Book*, John Wiley & Sons, 1994.

[VOL 99] VOLDMAN S.H., "The state of the art of electrostatic discharge protection: physics, technology, circuits, design, simulation, and scaling", *IEEE Journal of Solid-State Circuits*, vol. 34, no. 9, pp. 1272–1282, 1999.

[VOL 07] VOLDMAN S.H., *Latch-Up*, Wiley, 2007.

[VRI 07] VRIGNON B., LACRAMPE N., CAIGNET F., "Investigation of effects of an ESD pulse injected by a near-field probe into an oscillator block of a 16-bit microcontroller", *6th International Workshop on Electromagnetic Compatibility of Integrated Circuits (EMC Compo 2007)*, Turin, Italy, 28–30 November 2007.

[WAN 00] WANG Y., JULIANO P., JOSHI S. *et al.*, "Electrothermal modeling of ESD diodes in Bulk-Si and SOI technologies", *Electrical Overstress/Electrostatic Discharge Symposium Proceedings*, pp. 430–436, 2000.

[WAN 03] WANG K., POMMERENKE D., CHUNDRU R. *et al.*, "Numerical modeling of electrostatic discharge generators", *Electromagnetic Compatibility, IEEE Transactions on*, vol. 45, no. 2, pp. 258–271, 2003.

[WAN 04a] WANG K., POMMERENKE D., CHUNDRU R. *et al.*, "Characterization of human metal ESD reference discharge event and correlation of generator parameters to failure levels-part II: correlation of generator parameters to failure levels", *IEEE Transactions on Electromagnetic Compatibility*, vol. 46, no. 4, pp. 505–511, 2004.

[WAN 04b] WANG K., POMMERENKE D., "The PCB level ESD immunity study by using 3 dimension ESD scan system [C]", *International Symposium on Electromagnetic Compatibility*, pp. 343–348, 2004.

[WAN 06] WANG A.Z.H, *On-Chip ESD Protection for Integrated Circuits: An IC Design Perspective*, Springer Science & Business Media, 2006.

[WIL 86] WILSON T., SHEPPARD C.J.R., "Observations of dislocations and junction irregularities in bipolar transistors using the OBIC mode of the scanning optical microscope", *Solid-State Electronics*, vol. 29, no. 11, pp. 1189–1194, 1986.

[WIL 88] WILLS K.S., DUVVURY C., ADAMS O., "Photoemission testing for ESD failures. Advantage and limitations", *EOS/ESD 1988 5th Electrical Overstress/Electrostatic Discharge Symposium*, pp. 53–61, 1988.

[WOL 05] WOLF H., GIESER H., STADLER W. *et al.*, "Capacitively coupled transmission line pulsing cc-TLP – a traceable and reproducible stress method in the CDM-domain", *Microelectronics Reliability*, vol. 45, no. 2, pp. 279–285, 2005.

[WOR 01] Working group IEEE: "Cabling ESD study", *IEEE P802.3 – Static Discharge in Copper Cable Ad-hoc*, available at: http://www.ieee802.org/3/ad_hoc/copperdis/public/docs/index.html, March 2001.

[WUN 68] WUNSCH D.C., BELL R.R., "Determination of threshold failure levels of semiconductor diodes and transistors due to pulse voltages", *IEEE Transactions on Nuclear Science*, vol. 15, no. 6, pp. 244–259, 1968.

[XIA 03] XIAOFEN R., XIJUN Z., ZHANCHENG W. *et al.*, "Study on two types of commercial ESD simulators", *Conference on Environmental Electromagnetics*, pp. 233–236, 2003.

[ZAK 92] ZAKY S.G., BALMAIN K.G., DUBOIS G.R., "Susceptibility mapping", *International Symposium on Electromagnetic Compatibility*, pp. 439–442, 1992.

Index

1 Ω method, 39–41

A, B, C

ageing, 183, 188
antistatic, 75, 76
attenuation coefficient, 28, 38, 48
avalanche multiplication coefficient, 83, 130, 142
ballast resistance, 85, 90, 112, 123, 126, 138
behavioral modeling, 142, 159, 160, 163, 166, 196, 197, 209
bi-directional protection, 101
breakdown voltage, 72–74, 76, 81–83, 92, 93, 96, 116, 117, 128, 130, 134, 137, 142, 185, 189
Cable Discharge Event (CDE), 7, 17, 18
calibration, 8, 10, 47, 49, 52, 115, 116, 119, 208
Capacitively Coupled TLP (CC-TLP), 25
carrier lifetime, 118
central protection, 93, 101, 135, 157, 177, 180

characteristic impedance, 26, 29–31, 35–37, 46
Charged Board Model (CBM), 59
Charged Device Model (CDM), 4
compact modeling, 111, 126, 127, 138–140, 156–158, 166, 175
conductivity modulation, 127, 131
cut-off frequency, 146, 147

D, E, F

diaphonic coupling, 36
distributed protection, 79, 97
DPI method, 56, 57
DTSCR, 96
dynamic
    light emission microscopy (PICA), 60–62
    resistance, 154, 160
Electrical Fast Transient (EFT), 8
Electromagnetic Compatibility (EMC), 13, 21, 57
Electrostatic Discharges (ESD), 1, 71, 182, 184

Printed in the United States
By Bookmasters